1 .... HANDBOOK

Fourth Edition

DAVID SOYE

**KOGAN
PAGE**

YOURS TO HAVE AND TO HOLD
BUT NOT TO COPY

First published in 1985
Second edition 1990
Third edition 1995
Fourth edition 1997

Apart from any fair dealing for the purposes of research or private study, or criticism or review, as permitted under the Copyright, Designs and Patents Act, 1988, this publication may only be reproduced, stored or transmitted, in any form or by any means, with the prior permission in writing of the publishers, or in the case of reprographic reproduction in accordance with the terms and licences issued by the CLA. Enquiries concerning reproduction outside those terms should be sent to the publishers at the undermentioned address:

Kogan Page Limited
120 Pentonville Road
London N1 9JN

© David Soye, 1985, 1990, 1995, 1997

British Library Cataloguing in Publication Data
A CIP record for this book is available from the British Library
ISBN 0 7494 2396 X

Typeset by BookEns Ltd, Royston, Herts.
Printed and bound in Great Britain by Drogher Press, Christchurch, Dorset.

Contents

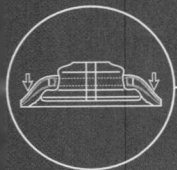

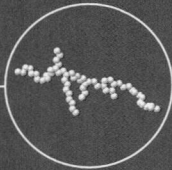

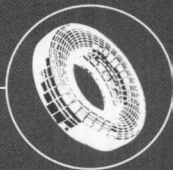

Acknowledgements

The author wishes to formally acknowledge the following people who have so kindly offered their help and assistance in the preparation of this edition of the *Truck Driver's Handbook*.

All road signs and markings appear with the kind permission of the controller of Her Majesty's Stationery Office.

The tachograph illustrations come courtesy of Smith Industries and Lucas Kienzle, reproduced from the *Transport Manager's Guide to the Tachograph* by David Soye, published by Kogan Page.

Also to Lion Industries who gave their permission to reproduce illustrations and text of the Intoximeter, and the Metropolitan Police for their advice with regard to the procedures followed in the event of a driver being charged with the offence of drinking and driving.

The First Aid illustrations and text appears with the kind permission of Kays Medical.

Thanks to the Management and Staff of Le Shuttle-EuroTunnel for their contribution towards the chapter relating to the Channel Tunnel.

Essential information was compiled with the assistance of the Road Haulage Association (RHA) and the *ABC Freight Guide* (Centaur Publications).

Finally, to Manchester Training who offered their help and assistance in compiling the essential information that has contributed in the preparation of this fourth edition of the *Truck Driver's Handbook*.

Introduction

Drivers of large, heavy vehicles have always been required to be familiar with the various rules and regulations affecting the vehicle and its load. In the past, this knowledge was relatively simple and slow to change but in today's world of transport the rules and regulations are complex and change frequently. As transport and distribution moves into the next century truck drivers need to be more and more knowledgeable and professional in their approach to the job in hand.

The advancement in vehicle technology is surpassed only by the ever increasing legislation affecting the vehicle and load, and the numerous rules and regulations regarding licensing, speed limits, driver's hours, tachographs, and so on.

People still think of this industry as being largely composed of 38 ton juggernauts travelling on long journeys every day of the week, when in fact only a small percentage of trucks fall into this category. This means that the majority of truck drivers are not engaged in this type of work but have jobs which involve a lot more than just driving. Most drivers are responsible for loading many kinds of goods, varying in quantities, size and condition, protecting the load against the weather and other hazards and delivering it safely, promptly and without fuss while, at the same time, maintaining front-line customer relations on behalf of their company.

Although a driver is not expected to be an engineer or hold a degree in road transport law, he does need to be conversant with the rules of the road and the many requirements and restrictions which affect the day-to-day duties of the job.

Professional truck driving is a demanding occupation but the compensations are many. The combination of freedom and responsibility holds a great attraction for people with skill and ability. Financial rewards vary widely according to experience, knowledge, expertise and the degree of responsibility required in each particular set of circumstances. Those drivers who enjoy the highest reward have usually graduated to these heights through hard work, the ability to learn and the intelligent application of common sense in all situations.

Although the *Truck Driver's Handbook* provides a ready-made source of transport data for experienced drivers, it also provides a mass of essential information for the driver who is entering the industry for the first time.

The section dealing with load security will be of immense value to the newcomer as it takes him step-by-step through each stage in the art of safe loading, offering a few practical hints on interlocking cargoes, loading drums and barrels, in addition to the safe loading of sheet steel or bars using correctly positioned skids including the art of roping and sheeting. It explains in simple terms, supported with easy-to-follow illustrations, major rope joins, spreaders, dolly knots and other useful ways in which the driver may use the rope to secure the load to the vehicle. Lorry sheets are usually big, bulky and heavy and can often be difficult to handle. The book will explain how to handle the sheets and get the best use from them, how to fold them, and how to cover and protect loads using one, two or even three sheets.

Experienced and new drivers will benefit from the information being offered concerning wide loads and long loads, maximum vehicle lengths, weight and height limits. There is a special feature on the drinking and driving laws, which take a driver through each stage of this very serious offence from the moment he is required to take the preliminary breath test at the side of the road, to how it could affect his insurance if convicted.

The LGV driver's licence is explained in detail – how to obtain it, how to renew or exchange it, and how easily it can be taken away through serious driving offences, such as drinking and driving, dangerous driving or penalty point accumulation. Many other driving offences are also covered in detail including the various speed restrictions imposed either by the type of vehicle being driven or by the type of road on which the vehicle happens to be at the time.

Other sections included in the *Handbook* cover the National Vocational Qualifications (NVQ). Open to all truck drivers and warehousing staff, they undoubtedly enhance any career prospects.

Drivers' hours and records and the correct use of the tachograph have been included to help the driver understand these complex rules.

The carriage of dangerous goods has become a specialised operation strictly controlled by detailed legislation. A section has been included in this new edition that will explain the ADR requirements and where the driver can go to be trained in order to get the ADR Certificate Entitlement.

Special features including the Channel Tunnel and driving overseas have been included to give the driver an insight into the development of transport as it takes him/her step-by-step through the terminal.

Other features include First Aid, coupling and uncoupling of articulated vehicles, demountable bodies, skip loaders and draw-bar trailers.

Many of the features outlined in the *Handbook* have a direct or indirect bearing on the duties of a truck driver and it cannot be emphasised enough that there are no substitutes for practical experience and associated knowledge. The *Truck Driver's Handbook* will go a long way in furnishing drivers with essential knowledge and information, but should a driver find himself in any doubt whatsoever he must not hesitate to *ask* for advice.

IMPORTANT NOTE

The information contained in this book is intended only as a layman's guide to legal requirements. It is not a definitive interpretation of the law. Only the courts are empowered to finally determine precisely what the law means. For this reason operators who encounter difficulties with law are advised to seek expert legal guidance.

1
The Large Goods Vehicle Driving Licence

APPLICATION FOR A DRIVING LICENCE

In order to drive a motor vehicle anywhere in the United Kingdom or in any member state of the European Union (EU), a driver must first have the correct licence.

In order to obtain an LGV driver's licence or provisional entitlement to drive a large goods vehicle, the applicant must first:

- hold a full driving licence for category **B** vehicles
- be 21 years of age or over
- be medically fit.

D1

Application for a driving licence

This form covers ALL driving licence applications.
If you need more information, see leaflet D100 available from Post Offices.

How much will it cost?

First provisional licence (car, motorcycle, medium/large goods, minibus/bus) (You can get a Highway Code from HMSO and other major bookshops)	£21
Changing provisional for first full	FREE
Duplicate If your licence is lost, stolen, destroyed or defaced.	£6
Exchange This includes adding a test pass to a full licence, adding provisional motorcycle entitlement, removing entitlements, or exchanging an old style pink or green licence for a new style one.	£6

Renewing your licence

Car licence	£6
Medium goods/large goods, minibus/bus	£21
For medical reasons	FREE

Exchanging licences from other countries

Full Northern Ireland car licence	£6
Full Northern Ireland licence for medium/large goods, minibus/bus	£21
Full EC/EEA or other foreign licence (including Channel Islands and Isle of Man).	£21

New licence after disqualification

Car licence, motorcycle, medium/large goods, minibus/bus	£12
All provisional licences or full Medium/large goods or minibus/bus licences with less than 3 months to run	£21
If disqualified for some drink/driving offences (see leaflet D100)	£20

If you are disqualified, and have to take another driving test, you normally have to pay £12 for a provisional licence. You also have to pay £6 for a full licence after passing the test.

How to pay
Please do not send us cash, bank notes or blank postal orders. We cannot be responsible if they go missing. Make out your cheque or postal orders to: 'Department of Transport' and cross them 'Drivers Account'. Please do not postdate your cheque - we cannot accept it.

On the back of your cheque
Write your full name, address, date of birth or driver number if known.

Where to send the form

Send the form to:	DVLA, SWANSEA
Use the correct postcode:	
First provisional licence	SA99 1AD
First, full licence for cars and motorbikes	SA99 1BJ
Other licences to drive cars and motorbikes	SA99 1AB
All medium/large goods, minibus/bus licences	SA99 1BR

Organ donation
If you wish to donate an organ to help someone else after your death, please complete the voluntary organ donor consent section in 'Your details' box. If you want more information, freephone 0800-555-777.

When you have sent off the form

It usually takes about three weeks from the day you post your form until your licence arrives. It might take longer if we have to check on your health. If your licence has not arrived in this time, please write to the:

Customer Enquiries Unit, DVLA, Swansea SA6 7JL.

or phone them on 01792-772151 or fax them on 01792-783071 (Please do not send your application to this address - see the above note "Where to send the form").

If you need to use microcom because of your hearing, ring 01792-782787 for driving licence enquiries.

You will need to tell them your driver number or your full name and date of birth.

Ring between 8.15 am and 4.30 pm Monday to Friday.

When you telephone you will be greeted by our Interactive Voice System. However, during office hours callers will also be given the option of speaking to an operator. Please note some calls are monitored for quality purposes.

If you want to ask about vehicle licensing, ring 01792-772134, or fax them on 01792-782793.

For overseas users, ring 01792-782758 for vehicle enquiries.

Driving while we have your form

If you are applying for your first provisional licence, you must not drive until the licence arrives and comes into effect.

Other licences
Once we have received your valid application, the law allows you to drive before the licence arrives as long as:

- you have held a licence before
- you are not disqualified from driving (there are different rules for medium/large goods, minibus/bus drivers - please see leaflet D100)
- you have not been, and would not be, refused a licence for medical reasons
- you keep to any special conditions which apply to the licence

Your driver number
Please make a note here of all of your driver number if you have one in case you need to contact DVLA.

What is the serial number of your test pass certificate (if known)

The address on your licence
The format of the address which will be printed on your licence is that authorised by the Post Office. It may differ from the address you supply on the application form. You must tell us when you change your name and/or address.

Application form for a driving licence

These notes are to help you fill in the form. It is not a statement of law.
There is more information in leaflet D100 which you can get from a Post Office, Traffic Area Office, Vehicle Registration Office or from DVLA.

Driving cars and motorbikes

If you are applying to drive cars or motorbikes only fill in the green parts of the form.

This will also enable you to drive mopeds, tractors, road rollers and some other vehicles. Please look in leaflet D100 for a full list.

Provisional licences

You need a provisional licence while learning to drive or ride.

Motorbikes

This includes scooters, and motorbikes with side cars. It doesn't include mopeds - so just tick 'without motorbikes' if you want to ride a moped.

If you are 16 you can only ride a moped. If you will want to ride motorbikes from age 17, tick 'with motorbikes' and your motorbike and provisional car entitlement will start from your 17th birthday.

You must complete Compulsory Basic Training before you can ride on the road. This also applies to mopeds. You can get details of training courses from the Driving Standards Agency - ring them on 0115-901 2500.

Full licences

If you passed your test on or after 1 April 1991 you must apply for a full licence within 2 years of passing the test.

If your test was passed before 1 April 1991 you must apply for a full licence within 10 years of passing the test.

If you leave it longer than this you will need to take another test.

Exchanging your licence

You can get an exchange licence if you want us to:

* add new categories to your full licence - please send us your test pass certificate and your licence
* remove out of date endorsements (see leaflet D100)
* add or take off provisional motorbike entitlement
* exchange an old style pink or green licence for a new style one

Please send us your old licence.

Licence from other countries

You can exchange a full Northern Ireland licence or test pass for a GB licence, or you can use your Northern Ireland licence here until it runs out.

EC/EEA countries

If you hold a valid full licence issued in any EC/European Economic Area (EEA) country, you need not exchange it for a GB one. You may drive here until the licence expires or until you reach the age at which British renewal becomes necessary, (see leaflet D100). However, you may if you wish, apply to exchange your licence for a British one at any time even if your licence has expired. If your EC/EEA licence was issued in exchange for one from another country, this may be valid in GB for only 12 months and you may not be able to exchange it for the GB equivalent.

Non EC countries

If you hold a full licence from a non EC country or Gibraltar you can use your licence here for one year provided it is valid. If you hold a valid full licence from a designated country (see leaflet D100), you can exchange it for a GB licence up to five years after coming to live here. If your licence was issued by a non EC country it may be possible for us to return it to you. Make your request at the time you apply to exchange.

Test passes from Gibraltar may be exchanged.

Licences issued by the Channel Islands or Isle of Man authorities may be exchanged provided they have been valid within the last 10 years.

No other test passes are acceptable.

For further advice, please contact the Customer Enquiries Unit (details on front page).

Your health

Please consult your doctor or optician if you are in any doubt whether you should declare a medical condition.

Residence

Normal residence means the place where you have personal or occupational ties. However, if you have moved to the UK having recently been permanently resident in another state of the EC/EEA, you must have been normally resident in the UK for 185 days in the 12 months prior to your application for a full driving licence.

Driving medium and large goods vehicles, minibuses and buses

If you are applying to drive these vehicles fill in both the green and blue parts of the form.

Provisional licences

A provisional licence for category C1 allows you to learn to drive medium sized goods vehicles, (between 3500 kgs and 7500 kgs). A provisional licence with category C allows you to learn to drive large goods vehicles (over 3500 kgs). A provisional licence with category D1, allows you to learn to drive minibuses (with up to 16 passenger seats) and provisional category C will allow you to learn to drive buses (with more than 8 passenger seats).

You MUST have a full car licence before learning to drive any of these vehicles and you MUST pass a driving test in a right vehicle before you can apply for entitlement to tow a trailer.

Please refer to leaflet D100 for a full description of the categories and minimum ages.

If you are in the armed forces and you want provisional vocational entitlement for all categories tick the appropriate box.

If this is your first application for provisional category C1, C, D1 or D, you must not drive that category of vehicle until you get the licence.

Full licences

If you have passed your test since 1 April 1991, you must apply for a full licence within 2 years of passing the test. If you leave it longer than this you will need to take another test.

Exchanging your licence

You can exchange your licence if:

* you want us to add new categories to your full GB licence -
  please send us your test pass certificate and your licence
* you want us to remove out of date suspension details

Licences from other countries

You can exchange a full Northern Ireland licence or test pass for a GB licence, or you may use your Northern Ireland licence here until it runs out.

EC/EEA countries

If you hold a valid full licence issued in any EC/European Economic Area (EEA) country, the same rules apply as those for car licence holders.

You can also exchange your licence for a full GB one if in the last 10 years you have held a full licence in Jersey or the Isle of Man.

Non EC countries

If your licence is from any other country, please contact the Customer Enquiries Unit (details on front page) for further information.

Your health

There are stricter rules about health for medium/large goods, minibus/bus drivers.
(see leaflet D100)

Please consult your doctor or optician if you are in any doubt whether you should declare a medical condition.

If this is your first application you must get a doctor to fill in a medical report on form D4 and ensure that all the relevant questions are completed.

You will also need to send a D4 if you are renewing your licence and you are 45 or over. If you have sent us a medical report in the last 12 months or if you hold a full EC/EEA licence and are under 45, do not send one now.

Duplicate licences

If your licence has been lost, stolen or destroyed you should apply for a duplicate using this form. Alternatively, provided none of the details on your licence has changed or is incorrect, you can apply for a duplicate licence by telephone using a credit or debit card. We accept Access, Visa, Eurocard, Mastercard and Switch. To use this service ring 01792-772151 between 8.15am and 4.15pm Monday to Friday. If you find your licence after requesting a duplicate you should return the original licence to the Agency with an explanatory note.

Please tear off this page and keep it - just send us the APPLICATION

Application form for a driving licence (cont.)

D: APPLICATION - Everyone must fill in the green parts. Lorry and bus drivers must fill in the blue parts as well.

1 Your details — Give your postcode address in England, Scotland or Wales and note "D is address on your licence"?

Surname

First names

Mr ☐   Mrs ☐   Miss ☐   Ms ☐   Other (e.g. Rev.) ☐

Title

Male ☐   Female ☐

What is your date of birth?   Day   Month   Year

What is your driver number? (if you have one)

Address

Post town   Postcode

Full daytime phone no.   What is your country of birth?

Organ Donation (Voluntary). I wish to donate an organ to help someone else after my death. Please register me on the NHS Organ Donor Register as someone whose organs can be used for transplants (Please tick)
A. Any part of my body ☐   or my   B. Kidneys ☐   C. Corneas ☐
D. Heart ☐   E. Lungs ☐   F. Liver ☐   G. Pancreas ☐

2 Your eyesight and your hearing
a. Can you read (with or without glasses or corrective lenses) a car number plate from 20.5 metres (67 feet)?   Yes ☐   No ☐
b. Do you need to wear glasses/corrective lenses when driving?   No ☐   Yes ☐
c. Do you need to wear a hearing aid or to use any other device to aid communication when driving lorries and buses?   No ☐   Yes ☐

3 The licence you want (see note on reverse)
Car/motorbike licences
Duplicate licence
My licence has been > Lost ☐   Stolen ☐   Destroyed ☐   Defaced ☐   (Please still send it to us)

First Provisional
☐ without motorbikes
☐ with motorbikes
If you are 16 and are getting Disability Living Allowance (mobility allowance) at the higher rate, and want to drive a car, tick box ☐

Exchange my licence to:
☐ add categories to my licence
  Enter the category passed in the box ☐
☐ remove out of date endorsements
☐ add/take off provisional motorbike entitlement
☐ exchange my pink/green licence for a new one

Renew my licence:
☐ at age 70 or over
☐ as it was withdrawn for medical reasons
☐ as I was disqualified

Exchange my Northern Ireland or foreign licence
What country is it from? ☐

Your licence will begin from the day we issue it. If you want to start later, when do you want it to start? (This cannot be more than 2 months after the day you apply, or 3 months after for lorry and bus drivers).

All applicants MUST complete this section

Medium/large goods vehicles, Minibus/bus licences

Provisional (please specify)
Goods vehicles
☐ (C1)   ☐ (C1+E)   ☐ (C)
passenger carrying vehicles
☐ (D1)   ☐ (D)
Are you in the armed forces   No ☐   Yes ☐
If you are and you want all provisional vocational entitlement please tick ☐
Are you in the Young Drivers' training Scheme   No ☐   Yes ☐
Full as I have passed my test
☐ add categories to my licence
Enter the category passed in the box ☐
(Remember to enclose your D10V test pass certificate with your application)

Renew my licence:
☐ at present one is due for renewal
☐ as it was withdrawn for medical reasons
☐ as I was disqualified

Exchange my licence for:
☐ take off out of date suspension details

Exchange my Northern Ireland or foreign licence
What country is it from? ☐

4 Previous licence details
What type was your licence?
tick appropriate box   FULL ☐   PROV ☐
Are you currently disqualified from driving in any other EC/EEA country   No ☐   Yes ☐
If yes, which country disqualified you ☐
If you have given up driving because you were disqualified in GB, please give date of disqualification ☐
Which court dealt with it? ☐

If you have not held a licence before, please tick this box ☐
You must send your old licence to us if you still have it.
What is the expiry date shown on the licence, or entitlement you are renewing? ☐
Has your name (including maiden name if applicable) or address changed since your last licence?   No ☐   Yes ☐
If Yes, fill in the old details below ☐

For Official use

TRANS TYPE

ISSUE No

FEE

REM TYPE

M/C

O/D

S/C

D.O.C.

T/P

D. A. M.

H/C

Please turn over ☞

Application form for a driving licence (cont.)

Your health

If you are sending in medical report form D4 with your medium goods/large goods, minibus or bus application, go on to Section 6. IF NOT, YOU MUST COMPLETE EITHER PART A OR PART B, OTHERWISE YOUR APPLICATION WILL BE SENT BACK TO YOU. If you have already told us about a medical condition that could affect your fitness to drive - and you have no new medical condition - miss out Part A and go on to Part B of this section.

PART A — HAVE YOU EVER HAD, OR DO YOU AT PRESENT SUFFER FROM, ANY OF THESE CONDITIONS?
(please answer ALL 17 questions)

1 An epileptic event (seizure or fit) — No ☐ Yes ☐
2 Sudden attacks of dizziness, giddiness, fainting or blackouts — No ☐ Yes ☐
3 Severe mental handicap — No ☐ Yes ☐
4 A pacemaker, defibrillator or antiventricular tachycardia device fitted — No ☐ Yes ☐
5 Diabetes controlled by insulin — No ☐ Yes ☐
6 Diabetes controlled by tablets — No ☐ Yes ☐
7 Angina (heart pain) while driving — No ☐ Yes ☐
8 A major or minor stroke — No ☐ Yes ☐
9 Parkinson's disease — No ☐ Yes ☐
10 Any other chronic neurological condition — No ☐ Yes ☐

11 A serious problem with memory — No ☐ Yes ☐
12 Serious episodes of confusion — No ☐ Yes ☐
13 Any type of brain surgery, brain tumour or severe head injury involving hospital inpatient treatment — No ☐ Yes ☐
  If Yes, give date:
14 Any severe psychiatric illness or mental disorder — No ☐ Yes ☐
15 Continuing/permanent difficulty in the use of your arms or legs which affects your ability to control your vehicle safely — No ☐ Yes ☐
16 Have you been dependent on or misused alcohol, illicit drugs or chemical substances in the past 3 years (this not include over/disorder offences) — No ☐ Yes ☐
17 Any visual disability which affects BOTH eyes (do not include short/long sightedness or colour blindness) — No ☐ Yes ☐
  If Yes, what is it?

PART A — If you are applying for medium goods/large goods, minibus/bus entitlement, you MUST answer these four extra questions.
Have you ever had any of these?
1 Sight in only one eye — No ☐ Yes ☐
2 Any visual problems affecting either eye — No ☐ Yes ☐
3 Angina — No ☐ Yes ☐
4 Any heart condition or heart operation — No ☐ Yes ☐
  If yes to 3 or 4, give dates and details

PART B — Only fill this in if you have told us about a medical condition before.
What is the condition? _____ Has it got worse since you told us about it? No ☐ Yes ☐
Have you had any special controls fitted to your vehicle since your last licence was issued? No ☐ Yes ☐

6 Convictions

Only fill in this part if you are applying for large goods and/or minibus/bus entitlement. Applicants for large goods entitlement should give details below of driving offences relating to drivers' hours or records, or road worthiness or loading of vehicles. Applicants for minibus/bus entitlement should also declare any convictions for non-driving offences.

Date of conviction	Court	Offence	Sentence or fine

Use a separate sheet if need to and attach to this form. Look in leaflet D-DD if you are not sure what to tell us about.

7 Declaration — If you or anyone else gives false information to help you get a driving licence, you or they can be prosecuted.

Residence: We cannot issue you with a full licence unless either: (i) you are normally resident in this country or (ii) you have been attending a course of study in this country for at least 6 months or (iii) you hold a full licence issued in Northern Ireland [see notes on residence]

What is the serial number on your test pass certificate? _____
(Remember to enclose it if you are claiming new categories)

Your fee Check the amount in the notes - DO NOT SEND CASH
Amount £ _____ Cheque or postal order number(s) _____

Have you enclosed:
☐ your last licence or licences? (if applicable)
☐ your fee? (remember to sign your cheque)
☐ your test pass certificate? (if applicable)
☐ a medical report form D4? (if needed)

Please write name and date of birth on the back of cheque/postal orders. This declaration MUST be signed by the person applying for the licence. I declare that I have checked the details I have given and to the best of my knowledge they are correct and I am entitled to the licence for which I apply.

Sign here _____   Date _____

Please tear off and keep the notes page - just send us the APPLICATION

Application form for a driving licence (cont.)

When making an application to have LGV provisional entitlement added to an ordinary licence the driver will need to make the application on a **D1 Application for a driving licence**. This form covers all driving licence applications including large goods vehicle and passenger carrying vehicle. The form can be obtained from a general post office, Traffic Area Office or any professional driver training centre.

In addition to the licence application form, the driver will also require a **D4 Medical Report Form.** This form is required on an application for large goods vehicle and/or passenger carrying vehicle. Also available from the Post Office, etc.

When completing the form D1 be sure to answer all the questions accurately and refer to the Guidance Notes attached for assistance.

The medical form D4 must be completed by a doctor after a routine medical examination.

Note: The medical examination is not available on the National Health and applicants must pay the examination fee themselves.

When both forms have been completed the applicant should send them to:
DVLC Swansea,
SA99 1BR
with the relevant licence fee.

Note: A driver must not drive any form of large goods vehicle until such time as the licence has been returned identifying the appropriate LGV entitlement.

MEDICAL IN CONFIDENCE

Medical Report

D4
Rev. 11/97

Medical Report on an applicant for Medium/Large Goods or Passenger Carrying Vehicle entitlement

- If this is your first application for the above you MUST send in this Medical Report form D4 completed by a Doctor. You must also do this if you are applying to renew your licence on or after age 45 and every 5 years until age 65. (If you are aged between 45-65 and have been issued a medical short period licence then you only need to submit a D4 with the renewal which falls nearest to the 5 year period since you last submitted one).

  From age 65 licences are issued for one year and each renewal MUST be accompanied by a D4.

- EC/EEA licence holders whose authority to drive in Great Britain has expired also need to have this form completed by a doctor in support of their application for a British licence. Further details about this can be found in form D100 available at Post offices.

A WHAT YOU HAVE TO DO

1. BEFORE consulting your Doctor please read the notes overleaf at Section C, paragraphs 1, 2 and 3. ("Medical standards for drivers of medium/large goods or passenger vehicles"). If you have any of these conditions you will NOT be granted this entitlement.

2. If, after reading the notes, you have any doubts about your ability to meet the medical or eyesight standards, consult your Doctor/Optician BEFORE you arrange for this medical form to be completed. The Doctor will normally charge you for completing it. In the event of your application being refused, the fee you pay the Doctor is NOT refundable. DVLA have NO responsibility for the fee payable to the Doctor.

3. Fill in Section 9 AND Section 10 on page 8 of this report in the presence of the Doctor carrying out the examination.

4. This report, together with your application, must be received at DVLA within 4 months of the Doctor signing the report. Failure to submit both forms together will lead to difficulties and delay in processing your application.

5. Please remove pages 1 and 2 before sending in the form with your application and check that all the sections have been completed fully.

B WHAT THE DOCTOR HAS TO DO

1. Please complete sections 1-8 of this report. You may find it helpful to consult the DVLA's "At a Glance" and the Medical Commission on Accident Prevention booklet - "Medical Aspects of Fitness to Drive". Further help can be obtained by telephoning 01792-783686 and asking to speak to one of the Medical Advisers. We would need to know the applicant's full name, address and date of birth. After hours there is an answerphone and we would need to know in addition your name, the surgery address and phone number and a time when it might be convenient to return your call.

2. Applicants who may be asymptomatic at the time of the examination should be advised that, if in future they develop symptoms of a condition which could affect safe driving and they hold any type of driving licence, they must inform the Drivers Medical Group, D7, DVLA, Swansea. SA99 1TU - immediately.

3. PLEASE ENSURE THAT YOU HAVE COMPLETED ALL THE SECTIONS

IF THIS REPORT DOES NOT BRING OUT IMPORTANT CLINICAL DETAILS WITH RESPECT TO DRIVING, PLEASE GIVE DETAILS IN SECTION 7.

Medical Report form

C   MEDICAL STANDARDS FOR DRIVERS OF MEDIUM/LARGE GOODS OR PASSENGER CARRYING VEHICLES

Medical standards for drivers of large vehicles in categories C1, C1+E, D1, D1+E, C, C+E, D and D+E are higher than those required for car drivers.

The following conditions are a bar to the holding of any of these entitlements.

1.   EPILEPSY ATTACKS

Applicants must NOT "have a liability to epileptic seizures". (this means that applicants must have been free of epileptic seizures for at least the last ten years and have not taken anti epileptic medication during this ten year period). With such a liability DVLA must refuse or revoke the licence.

2.   DIABETES

Insulin treated diabetics may NOT obtain a licence UNLESS they held a HGV/PSV licence valid at 1 April 1991 and the Traffic Commissioner in whose area they lived, or who issued the licence, had knowledge of the insulin treatment before 1 January 1991.

3.   EYESIGHT

All applicants, for whatever category of vehicle, must be able to read in good daylight a number plate at 20.5 metres (67 feet) and, if glasses or corrective lenses are required to do so, these must be worn while driving. In addition:

  (i) APPLICANTS FOR MEDIUM/LARGE GOODS OR PASSENGER CARRYING VEHICLE ENTITLEMENT MUST BY LAW HAVE:

    • A VISUAL ACUITY OF AT LEAST 6/9 IN THE BETTER EYE; AND

    • A VISUAL ACUITY OF AT LEAST 6/12 IN THE WORSE EYE; AND

    • IF THESE ARE ACHIEVED BY CORRECTION THE UNCORRECTED VISUAL ACUITY IN EACH EYE MUST BE NO LESS THAN 3/60.

An applicant who held a licence before 1 January 1997 and who has an uncorrected acuity of less than 3/60 in only one eye may be able to meet the required standard and should check with Drivers Medical Group, D7, DVLA, Swansea. SA99 1TU, or telephone 01792-783686, about the requirement.

An applicant who has held an LGV/PCV (formerly HGV/PSV) licence before 1.3.92 but who does not meet the standard in (i) above may still qualify for a licence. Information about the standard for such an applicant can be obtained from DRIVERS MEDICAL GROUP. (address as above).

  (ii) APPLICANTS ARE ALSO BARRED BY LAW FROM HOLDING MEDIUM/LARGE GOODS OR PASSENGER CARRYING VEHICLE ENTITLEMENT IF THEY HAVE:

    • UNCONTROLLED DIPLOPIA (DOUBLE VISION)

    • OR DO NOT HAVE A NORMAL BINOCULAR FIELD OF VISION

AN APPLICANT (OR EXISTING LICENCE HOLDER) FAILING TO MEET THE EPILEPSY, DIABETES OR EYESIGHT REGULATIONS MUST BE REFUSED BY LAW.

4.   OTHER MEDICAL CONDITIONS

In addition to those medical conditions covered by law, applicants (or licence holders) are likely to be refused if they are unable to meet the national recommended guidelines in the following cases:-

• within 3 months of myocardial infarction, any episode of unstable angina, CABG or coronary angioplasty

• a significant disturbance of cardiac rhythm occurring within the past 5 years unless special criteria are met

• suffering from or receiving medication for angina or heart failure

• Hypertension where the BP is persistently 180 systolic or over or 100 diastolic or over

• a stroke, TIA or unexplained loss of consciousness within the past 5 years

• Meniere's and other conditions causing disabling vertigo, within the past 1 year

• recent severe head injury with serious continuing after effects, or major brain surgery

• Parkinson's disease, multiple sclerosis or other "chronic" neurological disorders likely to affect limb power and co-ordination.

• suffering from a psychotic illness in the past 3 years, or suffering from dementia

• alcohol dependency or misuse, or continuing drug or substance misuse or dependency in the past 3 years

• incurable difficulty in communicating by telephone in an emergency

• any other serious medical condition which may cause problems for road safety when driving a Medium/Large Goods or Passenger Carrying Vehicle.

Medical Report form (cont.)

Medical Examination
to be completed by the Doctor *(please use black ink)*
Please answer all questions

D4
Rev. Jan/97

Please give patient's weight: _____ (kg/st and Height: _____ ft/cms

Please give details of smoking habits, if any: _____

SECTION 1 Vision (Please see EYESIGHT NOTES 3i to 3ii on page 2)

1. Is the visual acuity as measured by the Snellen chart AT LEAST 6/9 in the better eye and AT LEAST 6/12 in the other? (corrective lenses may be worn).   `YES` `NO`

2. Do corrective lenses have to be worn to achieve this standard?   `YES` `NO`

  (a) if YES, is the UNCORRECTED acuity AT LEAST 3/60 in the RIGHT eye?   `YES` `NO`

  (b) is the UNCORRECTED acuity AT LEAST 3/60 in the LEFT eye?   `YES` `NO`
    (3/60 being the ability to read the 60 line of the Snellen chart at 3 metres)

  (c) is the correction well tolerated?   `YES` `NO`

---

3. Please state all the visual acuities for all applicants:

UNCORRECTED		CORRECTED *(if applicable)*	
Right: _____	Left: _____	Right: _____	Left: _____

4. Is there a full binocular field of vision? (central and/or peripheral)
If NO, and there is a visual field defect please give details in SECTION 7
and enclose a copy of recent field charts, if possible.   `YES` `NO`

5. Is there uncontrolled diplopia?   `NO` `YES`

SECTION 2 Nervous System

                                      YES NO

1. Has the applicant had major or minor epileptic seizures?   ☐ ☐

  (a) If YES, please give date of last seizure _____

  (b) If treated, please give date when treatment ceased: _____

2. Is there a history of blackout or impaired consciousness within the last 5 years?   ☐ ☐
  (a) If YES, please give date(s) and details in SECTION 7

3. Is there a history of stroke or TIA within the past 5 years?   ☐ ☐
  (a) If YES, please give date(s) and details in SECTION 7

4. Is there a history of sudden disabling dizziness/vertigo within the last 1 year?   ☐ ☐
  (a) If YES, please give date(s) and details in SECTION 7

5. Does the patient have a pathological sleep disorder?   ☐ ☐
  (a) If YES, has it been controlled successfully?

6. Is there a history of chronic and/or progressive neurological disorder?   ☐ ☐
  (a) If YES, please give date(s) and details in SECTION 7

7. Is there a history of brain surgery?   ☐ ☐
  (a) If YES, please give date(s) and details in SECTION 7

8. Is there a history of serious head injury ?   ☐ ☐
  (a) If YES, please give date(s) and details in SECTION 7

9. Is there a history of brain tumour, either benign or malignant, primary or secondary?   ☐ ☐
  (a) If YES, please give date(s) and details in SECTION 7

APPLICANT'S NAME _____   DOB _____

Medical Report form (cont.)

Rev. 2/6/97

SECTION 3 Diabetes Mellitus

	YES	NO
1. Does the applicant have diabetes mellitus? *If YES, please answer the following questions. If NO, proceed to SECTION 4*	☐	☐
2. Is the diabetes managed by:-		
(a) Insulin?	☐	☐
(b) If YES, date started on insulin		
(c) Oral hypoglycaemic agents and diet?	☐	☐
(d) Diet only?	☐	☐
3. Is the diabetic control generally satisfactory?	☐	☐
4. Is there evidence of :-		
(a) Loss of visual field?	☐	☐
(b) Has there been bilateral laser treatment? If YES, please give date	☐	☐
(c) Severe peripheral neuropathy?	☐	☐
(d) Significant impairment of limb function or joint position sense?	☐	☐
(e) Significant episodes of hypoglycaemia?	☐	☐
(f) Complete loss of warning symptoms of hypoglycaemia?	☐	☐

SECTION 4 Psychiatric Illness

	YES	NO
5. Has the applicant suffered from or required treatment for a psychosis in the past 3 years? *(a) If YES, please give date(s) and details in SECTION 7*	☐	☐
6. Has the applicant required treatment for any other psychiatric disorder within the past 6 months? *(a) If YES, please give date(s) and details in SECTION 7*	☐	☐
7. Is there confirmed evidence of dementia?	☐	☐
8. Is there a history of alcohol misuse or alcohol dependency in the past 3 years?	☐	☐
9. Is there a history of continuing drug or substance misuse or dependency in the past 3 years? *If YES, to questions 8 or 9, please give details in SECTION 7*	☐	☐

SECTION 5 General

	YES	NO
1. Has the applicant currently a significant disability of the spine or limbs which is likely to impair control of the vehicle? *(a) If YES, please give details in SECTION 7*	☐	☐
2. Is there a history of bronchogenic or other malignant tumour with a significant liability to metastasise cerebrally? *(a) If YES, please give dates and diagnosis and state whether there is current evidence of dissemination*	☐	☐
3. Is the applicant profoundly deaf?	☐	☐
(a) If YES, could this be overcome by any means to allow a telephone to be used in an emergency?	☐	☐

APPLICANT'S NAME [ ]   DOB [ ]

Medical Report form (cont.)

Rev. Jan 97

SECTION 6 Cardiac

A. Coronary Artery Disease

	YES	NO
Is there a history of:		
1. Myocardial Infarction?	☐	☐
(a) If YES, please give date(s)		
2. Coronary artery by-pass graft?	☐	☐
(a) If YES, please give date(s)		
3. Coronary Angioplasty?	☐	☐
(a) If YES, please give date(s)		
4. Any other Coronary artery procedure? If YES, please give details in SECTION 7	☐	☐
5. Has the applicant suffered from Angina?	☐	☐
6. Is the applicant STILL suffering from Angina or only remains angina free by the use of medication?	☐	☐
7. Has the applicant suffered from Heart Failure?	☐	☐
8. Is the applicant STILL suffering from Heart Failure or only remains controlled by the use of medication?	☐	☐
9. Has a resting ECG been undertaken? If NO, proceed to question 12	☐	☐
(a) If YES, please give date		
10. Does it show pathological Q waves?	☐	☐
11. Does it show Left Bundle branch block?	☐	☐
12. Has an exercise ECG been undertaken (or planned)?	☐	☐
(a) If YES, please give date		
13. Has an angiogram been undertaken (or planned)?	☐	☐
(a) If YES, please give date and give details in SECTION 7		

B. Cardiac Arrhythmia

	YES	NO
1. Has the applicant had a significant documented disturbance of cardiac rhythm within the past 5 years? If YES, please give details in SECTION 7 If NO, proceed to section C overleaf	☐	☐
2. Has the arrhythmia (or its medication) caused symptoms of sudden dizziness or impairment of consciousness or any symptom likely to distract attention during driving within the past 2 years?	☐	☐
3. Has Echocardiography been undertaken? If YES, please give details in SECTION 7	☐	☐
4. Has an exercise test been undertaken? If YES, please give details in SECTION 7	☐	☐
5. Has a Cardiac defibrillator been implanted or anti-ventricular tachycardia device been fitted?	☐	☐
6. Has a PACEMAKER been implanted? If NO, proceed to SECTION C overleaf	☐	☐
7. If YES, was it implanted to prevent Bradycardia?	☐	☐
8. Is the applicant now free of sudden and/or disabling symptoms?	☐	☐
9. Does the applicant attend a pacemaker clinic regularly?	☐	☐

APPLICANT'S NAME _____ DOB _____

Page 6

Medical Report form (cont.)

C. Other Vascular Disorders

	YES	NO
1. Is there a history of Aortic aneurysm with a transverse diameter of 5cms or more? (Thoracic or abdominal) *If NO, proceed to SECTION D*	☐	☐
(a) If YES, has the aneurysm been successfully repaired?	☐	☐
2. Is there symptomatic peripheral arterial disease?	☐	☐
3. Has there been dissection of the Aorta?	☐	☐

D. Blood pressure

	YES	NO
1. Is there a history of hypertension with BP readings consistently greater than 180 systolic or 100 diastolic? *If NO, proceed to SECTION E* (a) If YES, please supply most recent readings with dates	☐	☐

_____

	YES	NO
2. If treated does the Medication cause any side effects likely to affect safe driving?	☐	☐

E. Valvular Heart Disease

	YES	NO
1. Is there a history of valvular heart disease (with or without surgery)? *If NO, proceed to SECTION F*	☐	☐
2. Is there any history of embolism?	☐	☐
3. Is there any history of arrhythmia - intermittent or persistent?	☐	☐
4. Is there persistent dilatation or hypertrophy of either ventricle? *If YES, please give details in SECTION 7*	☐	☐

F. Cardiomyopathy

	YES	NO
1. Is there established cardiomyopathy?	☐	☐
2. Has there been a heart or heart/lung transplant? *If YES, please give details in SECTION 7*	☐	☐

G. Congenital Heart Disorders

	YES	NO
1. Is there a congenital heart disorder? *If YES, please give details in SECTION 7*	☐	☐
2. If YES, is it currently regarded as minor?	☐	☐
3. Is the patient in the care of a Specialist clinic? *If YES, please give details in SECTION 7*	☐	☐

APPLICANT'S NAME _____ DOB _____

Medical Report form (cont.)

CONDITIONS AFFECTING LGV PROVISIONAL ENTITLEMENT

The holder of a licence with LGV provisional entitlement must comply with the following conditions when driving a Large Goods Vehicle:

- display 'L' plates clearly on the front and rear of the vehicle
- only drive under the supervision of a person over 21 years of age who holds a full UK LGV driving licence for the category of the vehicle being driven.

CLAIMING FULL LGV DRIVING LICENCE

In order to claim full LGV driving licence entitlement, the provisional licence holder will need to send to the licence authority the following:

- an LGV Driving Test Pass Certificate
- the Driving Licence signed on the reverse side.

On passing the LGV Driving Test, the LGV Entitlement will be included on the driver's licence. This Entitlement will be valid until the driver has reached the age of 45 years or for 5 years, whichever is the longer.

From the age of 45 LGV Entitlement is renewable every 5 years thereafter, up to the age of 65 when it becomes renewable every year.

Note: A medical report is required with each renewal from the age of 45. The same conditions apply to drivers holding Passenger Carrying Vehicles entitlement.

VEHICLE CATEGORIES AND ENTITLEMENT

In order to account for the varying degrees of skill in driving different categories of vehicles, it is necessary for a driver to progress from a small vehicle to a large vehicle.

Certain changes in the licence regulations which came into effect in January 1997 prevent a car driver from going straight to driving a 38 ton articulated vehicle without first passing the driving test on a rigid vehicle.

Category *Entitlement*		*Group*
A	Motor cycle with or without side car but excluding vehicles in category K or P. Additional Category B1 – P. *Note:* A provisional licence holder must not carry a pillion passenger even if that person is a qualified driver.	D A & B
B	Motor vehicle with a maximum authorised mass not exceeding 3.5 tonnes and not more than 8 seats in addition to the driver's seat, not included in any other category and including such a vehicle drawing a trailer not exceeding 750 kg. Additional Categories B + E, B1, C1 + E ,D, D1 + E, F, K, L, N, and P.	
B1	Motor tricycle with an unladen mass not exceeding 500 kg and with a maximum design speed exceeding 50 km per hour but excluding vehicles in categories K, L or P.	C

Category *Entitlement*		*Group*
C	Goods vehicle exceeding 3.5 tonnes mam* including such a vehicle drawing a trailer which, if it has a single axle, the weight of the trailer must not exceed 5 tonnes mam* or, in the case of any other trailer, 750 kg but not including any vehicle in category C1.	(LGV Classes 2 or 3)
C1	Goods vehicle exceeding 3.5 tonnes but not exceeding 7.5 tonnes mam*. Additional Categories B, B + E, B1, C1 + E, D1 + E, F, K, L, N and P.	
D	Passenger carrying vehicle having more than 8 seats in addition to the driver's seat including such a vehicle drawing a trailer which, if it has a single axle, the weight of the trailer must not exceed 5 tonnes mam* or, in the case of any other trailer, 750 kg, but not including any vehicle in category D1.	PCV Classes 1, 2 and 3
D1	Passenger carrying vehicle (*not* used for hire or reward) with more than 8 but not more than 16 seats, in addition to the driver's seat, and including such vehicles drawing a trailer not exceeding 750 Kg mam*. Additional categories B, B + E, B1, C1, C1 + E, D1 + E, F, K, L, N and P.	A or B
B + E	Combination of a motor vehicle in category B and a trailer with a mam* exceeding 750 kg.	A or B
C + E	Combination of a motor vehicle in category C and a trailer with a mam*, in the case of a trailer with a single axle, exceeding 5 tonnes, or 750 kg in any other case.	(LGV Class 1) (Articulated)

Category Entitlement		Group
C1 + E	Combination of a motor vehicle in category C1 and a trailer with mam* exceeding 750 kg mam* where the mam of the combination does not exceed 12 tonnes.	
D + E	Passenger carrying vehicle in category D and a trailer with a mam*, in the case of a trailer with a single axle, exceeding 5 tonnes, or 750 kg in any other case.	PCV Classes 1 and 2
D1 + E	A motor vehicle in category D1 and a trailer with a mam* exceeding 750 kg.	
F	Agricultural tractor but excluding any vehicle in category H. Additional category K.	F
G	Road Roller.	G
H	Track laying vehicle steered by its tracks.	K
K	Mowing machine or pedestrian controlled vehicle.	K
L	Electric vehicle. Additional category K.	
N	Vehicle exempted from duty under section 7 (1) of the Vehicle (Excise) Act 1971.	N
P	Moped.	E

* maximum authorised mass

COST AND DURATION OF A UNIFIED DRIVING LICENCE

Full unified licence	£21.00	Valid until holder's seventieth birthday
Provisional licence (not category A) that can be changed for a full licence after passing the relevant driving test	£21.00	Valid until holder's seventieth birthday
Exchange licence	£6.00	
Duplicate licence	£6.00	
Category A licence	£6.00	
Provisional Category A licence	£6.00	Valid for two years

Note: A fee of £6.00 is made to a person who requires a new licence to be issued after certain disqualifications. In the case of drinking and driving offences the fee is £20.00.

Production of a licence
If required to do so, a licence must be produced for examination by a police officer either on demand or within seven days at a specified police station.

Cost of a driving test
Car driving test, weekdays	£31.00
Car driving test, weekends	£41.00
LGV driving test, weekdays	£65.00
LGV driving test, weekends	£83.00
PCV driving test, weekdays	£65.00
PCV driving test, weekends	£83.00

Cost of theory test
All vehicles	£15.00

MINIMUM AGE LIMIT FOR DRIVERS

The minimum age requirement allows a driver of 17 years of age to drive small goods vehicles up to 3.5 tonnes; 18 years of age to drive light goods vehicles between 3.5 – 7.5 tonnes gross vehicle weight; 21 years of age to drive heavy/large goods vehicles over 7.5 tonnes gross vehicle weight, including the maximum permissible weight of any trailer drawn or semi-trailer in the case of articulated vehicles

Drivers with licence category **B** entitlement will not be allowed to tow a trailer where the maximum weight of the trailer exceeds 750 kgs without first passing the relevant driving test offering category **B + E** entitlement.

A vehicle weighing up to 2.5 tonnes could tow a trailer weighing up to 1 tonne without the need for the driver to hold Category **B + E** entitlement, because the joint weight of the vehicle and trailer does not exceed 3.5 tonnes.

However a small car or light van would not be allowed to tow a trailer or a caravan where the trailer or caravan exceeds the weight of the towing vehicle. In this case the driver would be required to pass the relevant driving test that would give category **B + E** entitlement.

The principles would allow a systematic progression of licencing covering the following examples:
1. Cars
2. Cars and caravans
3. Rigid goods vehicles
4. Rigid goods vehicles with trailers and articulated vehicles.

DRIVING OFFENCES, ENDORSEMENTS AND PENALTY POINTS

Where a driver has been convicted of an endorsable offence but has not been disqualified, he/she will incur penalty points which will be recorded on the unified licence irrespective of what category of vehicle was being driven at the time of the offence.

The accumulation of 12 penalty points or more within 3 years will result in the driver being disqualified.

Drivers convicted of more than one offence committed on the same occasion may incur separate penalty points for each separate offence.

Penalty points will normally remain on a licence for 3 years except in the case of offences relating to drink driving when the period is normally 11 years.

Offences	*Points*
Contravention of temporary prohibition or restriction	3 – 6
Use of special road contrary to scheme or regulations	3 – 6
Contravention of pedestrian crossing regulations	3
Not stopping at school crossing	3
Contravention of order relating to street playground	2
Exceeding speed limit	3 – 6
Causing death by dangerous driving	3 – 11
Dangerous driving	3 – 11
Careless and inconsiderate driving	3 – 9
Causing death by careless driving when under the influence of drink or drugs	3 – 11
Driving or attempting to drive when unfit to drive through drink or drugs	3 – 11
Being in charge of a vehicle when unfit to drive through drink or drugs	10
Being in charge of a mechanically propelled vehicle with excess alcohol in body	10

Failing to provide a specimen for breath test	4
Failing to provide specimen for analysis or laboratory test	3 – 11
Motor racing and speed trials on public ways	3 – 11
Leaving vehicles in dangerous positions	3
Failing to comply with traffic directions	3
Using vehicle in dangerous condition, etc	3
Breach of requirements as to brakes, steering, gear or tyres	3
Driving otherwise than in accordance with a licence	3 – 6
Driving with defective eyesight or refusing an eyesight test	3
Driving after making false declaration as to physical fitness	3 – 6
Failing to comply with conditions attached to a provisional or full licence	2
Driving following the failure to notify onset of, or deterioration in, relevant or prospective disability	3 – 6
Driving after refusal of licence or revocation	3 – 6
Driving while disqualified: where offender was disqualified as under age	2
where offender was disqualified by order of court	7
Using motor vehicle without insurance	6 – 8
Failing to stop after accident and give particulars or report accident	5 – 10
Failure of keeper of vehicle and others to give police information as to the identity of driver, etc in the case of certain offences	3
Manslaughter or, in Scotland, culpable homicide	3 – 11

2
LGV Driving Test

INTRODUCTION AND SUMMARY

The LGV driving test was first introduced into the United Kingdom in the early 70's, and there was very little change in the format, procedure or standard until 1 January 1997 when the test underwent significant changes in format and procedure.

The most significant changes in the test are:

1. The introduction of the new theory test.
2. The removal of the question and answer sessions on the Highway Code and the recognition of selected road signs.
3. The coupling and uncoupling of trailers and/or semi-trailers has become a practical exercise and drivers will no longer be required to answer questions on this as this is no longer a theoretical part of the test.

These changes along with licence changes

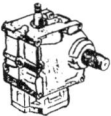

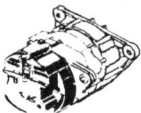

which came into effect in January 1997 mean that a driver **must first pass** the theory part of the LGV test before he/she can take the practical driving part of the test.

THE THEORY TEST

Drivers making an application to take the theory test will need to make their application on the appropriate 'Application for a Theory Test Appointment' form. These forms are available from driving test centres, the DSA area offices and most LGV driver training centres. The fee for the theory test is £15.00 and applies to all categories, including large goods vehicles, passenger carrying vehicles, motor cars and motor cycles with each category having its own special questions.

The theory test is a multiple-choice question paper which requires the selection of one correct answer from a choice of four. There are also some questions where two or three answers are to be selected from a choice of six. All questions must be answered correctly in the format described on the test paper.

The LGV and PCV theory test papers contain 25 questions each. Candidates will need to answer 21 questions correctly in order to pass the theory part of the test.

THEORY TEST EXAMPLE QUESTIONS

Select one answer unless otherwise indicated.

What is the maximum gross weight for an articulated vehicle with five or more axles?

(a)   36 tonnes ☐

(b) 38 tonnes ☐
(c) 40 tonnes ☐
(d) 45 tonnes ☐

The correct answer is (b).

Before driving over a level crossing, drivers of large goods vehicles will need to phone the signal operator if the gross vehicle weight is over:

(a) 30 tonnes ☐
(b) 32 tonnes ☐
(c) 35 tonnes ☐
(d) 38 tonnes ☐

The correct answer is (d).

What is the minimum height of an unmarked bridge?

(a) 4.5 metres (15 feet) ☐
(b) 4.7 metres (15 feet 6 inches) ☐
(c) 4.8 metres (16 feet) ☐
(d) 5 metres (16 feet 6 inches) ☐

The correct answer is (d).

What type of securing device would a driver use when carrying a load of steel?

(a) Straps ☐
(b) Chains ☐
(c) Ropes ☐
(d) Sheets ☐

The correct answer is (b).

How wide must a load be before the driver fits the red and white warning triangle?

(a) 2.6 metres (8 feet 6 inches) ☐
(b) 2.7 metres (8 feet 10 inches) ☐
(c) 2.8 metres (9 feet 2 inches) ☐
(d) 2.9 metres (9 feet 6 inches) ☐

The correct answer is (d).

What would you do if exhaust fumes were seeping into the vehicle?

(a) report it as soon as you return to the depot ☐

(b) stop and have the fault rectified ☐

(c) avoid excessive revving when the vehicle is stationary ☐

(d) have the emissions checked at the next vehicle inspection test ☐

The correct answer is (b).

On a three-line braking system what is the colour of the auxiliary line?

(a) Red ☐

(b) Blue ☐

(c) Green ☐

(d) Yellow ☐

The correct answer is (b).

When making entries on a tachograph chart which of the following should be made on the centre field?

(a) details of goods being carried ☐

(b) name and address of employer ☐

(c) the place at which you start your day's work ☐

(d) the rest period taken prior to starting the journey ☐

The correct answer is (c).

Drivers must return completed tachograph charts to their employers within:

(a) 21 days ☐

(b) 25 days ☐

(c) 28 days ☐

(d) 30 days ☐

The correct answer is (a).

During your break period, your vehicle will be moved by another driver. What should you do with the tachograph chart?

(a) remove the chart and make a manual record of the break period ☐

(b) leave the chart in the vehicle and record the change on the reverse side ☐

(c) put in a new chart on your return to the vehicle ☐

(d) switch to rest mode to record the break ☐

The correct answer is (a).

Under EU rules a driver must take a break after a continuous driving period of:

(a) 3 hours ☐
(b) 4 hours ☐
(c) 4.5 hours ☐
(d) 5.5 hours ☐

The correct answer is (c).

What action would you take if the brake pressure warning device started to operate while the vehicle was in motion?

(a) stop and get the fault rectified ☐
(b) drain the air tanks ☐
(c) disconnect the air lines ☐
(d) continue to drive the vehicle ☐

The correct answer is (a).

What is the main cause of brake fade?

(a) loss of air pressure in the system ☐
(b) badly worn brake pads ☐
(c) prolonged and continuous use of brakes ☐
(d) repeated pumping of the brakes ☐

The correct answer is (c).

What is the maximum speed limit for a large goods vehicle on a motorway?

(a) 50 mph ☐
(b) 55 mph ☐
(c) 60 mph ☐
(d) 70 mph ☐

The correct answer is (c).

After driving through water how would you dry the brakes?

(a) drive in low gear with the foot
     brake applied lightly ☐
(b) carry out an emergency stop ☐
(c) avoid braking until the brakes
     have dried out ☐
(d) pump the brake pedal periodically ☐

The correct answer is (a).

Vehicles fitted with audible warning devices when reversing should not be used in built-up areas:

(a) between 10.30 pm and 6.30 am ☐
(b) between 11 pm and 6.30 am ☐
(c) between 11.30 pm and 7 am ☐
(d) between 12.30 am and 8am ☐

The correct answer is (c).

Select three answers

When driving on a long journey what can you do to help keep you alert?

(a) eat a heavy meal before setting off ☐
(b) walk around in the fresh air after
     a rest period ☐
(c) keep the cab warm and comfortable ☐
(d) take proper rest periods at correct
     intervals ☐
(e) keep plenty of cool fresh air moving
     through the cab ☐

(f)   drive faster to get to the destination
      sooner ☐

The correct answers are (b), (d) and (e).

**Which of the following would make a tyre
illegal for a large goods vehicle?**

(a)   different makes of tyres on the
      same axle ☐
(b)   a tread depth of 1.3 mm ☐
(c)   exposed ply or cord ☐
(d)   a deep cut more than 25 mm (1 inch)
      long ☐
(e)   recut tyres ☐
(f)   a lump or bulge ☐

The correct answers are (c), (d) and (f).

Select two answers

**Which vehicles must be fitted with a
tachograph when operating in the UK under
EU rules?**

(a)   a light goods vehicle drawing a
      trailer exceeding 3.5 tonnes gross
      vehicle weight ☐
(b)   any vehicle over 3.5 tonnes gross
      vehicle weight ☐
(c)   vehicles between 3.5 and 7.5 tonnes
      gross vehicle weight ☐
(d)   vehicles over 7.5 tonnes gross
      vehicle weight used only for driving
      instruction ☐
(e)   vehicles manufactured before 1947 ☐
(f)   vehicles manufactured before 1957 ☐

The correct answers are (a), (b), (c) and (f).

THE PRACTICAL TEST

Applicants who have successfully completed

the theory test may then apply for the
practical test.

It is very important that the vehicle is
ready for the test, as well as the driver. The
driver must ensure that all the normal daily
vehicle checks are carried out and are
satisfactory prior to leaving base and driving
to the Test Centre.

The first part of the practical test is the
uncoupling and coupling of the trailer or
semi-trailer from the prime mover (C and E
only). This is a practical exercise where the
driver will be required to demonstrate his/her
skills in coupling and uncoupling.

The recognised procedures for uncoupling
and coupling of a semi-trailer to and from
the prime mover are outlined below.

FIFTH WHEEL COUPLING PROCEDURE

- Check that the trailer brake is on.
- Climb into the cab and reverse slowly,
  checking relative heights of tractor and
  trailer and making sure that the unit is in
  line with the trailer.
- Reverse until the coupling engages.
- Select a low forward gear and tug
  forward.
- When you are satisfied that the fifth
  wheel is connected, apply parking brake,
  climb down from the cab leaving it in a
  safe position.
- Do a visual check on the fifth wheel, and
  ensure that the safety catch is on.
- Climb up between the unit and trailer
  and connect airlines and electrical line
  (turn on taps if fitted).
- Raise landing gear and stow away the
  handle.
- Take off the trailer brake.

- Fix the number plate and carry out all other checks.
- Check lights and direction indicators.

FIFTH WHEEL UNCOUPLING PROCEDURE

- Make sure trailer is on firm and level ground.
- Apply the trailer brake.
- Lower the handling gear and stow the handle away.
- Climb up between the unit and the trailer.
- Turn off the air taps (if fitted).
- Disconnect the airlines and electrics, stow away.
- Climb down.
- Remove the safety catch on the fifth wheel and disconnect the wheel.
- Retrieve the number plate from the trailer, stow it in the cab.
- Get into the cab and drive away slowly, making sure that the trailer causes no damage as it settles.

REVERSING EXERCISE

This exercise will begin when the vehicle is in position at the bottom right hand corner of the manoeuvring area, whereupon the driving examiner will give clear instructions as to how the exercise is to be carried out. Great care must be taken when doing this exercise as there is very little margin for error. The driver who tends to oversteer, understeer or fails to position the vehicle *correctly* will undoubtedly bring pressure upon himself. The examinee must also avoid erratic or speedy reversing. The vehicle must be positioned correctly, using the accelerator, clutch and steering effectively. A slow but constant speed must be maintained, particularly with articulated vehicles. Ideally the candidate should be able to position the vehicle correctly and complete the exercise without having to stop or without having to take a shunt. If a driver is unable to do this, he should not be taking the test as clearly he is not ready.

BRAKING EXERCISE

The third part of the practical test is the vehicle controlled stop, or braking exercise. Here, the candidate will be directed to another part of the manoeuvring area where this exercise can be carried out safely. The candidate will be directed to the start line/ position where he will be required to park his vehicle in order to allow the driving examiner to climb in. Only when the examiner is settled in and has secured all his paperwork, will he give clear instructions as to how the exercise is to be carried out.

The driver will need to accelerate vigorously (subject to the power output of the vehicle) in order to attain a speed of 20–25 mph within the prescribed distance of 200 ft. When the front of the vehicle comes in line with the cones, the driver will be required to come off the accelerator and apply the footbrake firmly but without locking the wheels, keeping the footbrake depressed until the vehicle has come to an abrupt stop. The candidate must avoid locking the wheels, stalling the engine or in the case of articulated vehicles, jack-knifing the trailer.

Reversing Exercise

The exercise is commenced from a position with the front of the candidate's vehicle in line with marker cones A and A1. The candidate reverses into the bay, keeping marker B on the offside, and stops with the extreme rear of his vehicle within the 3 ft stopping area.

Distances   A – A1   = 1 1/2 times width of vehicle.
             A – B   = 2 times length of vehicle.
             B – Line Z = 3 times length of vehicle.

The width of the bay will be 1^1/$_2$ times the width of the vehicle. The length of the bay will be based on the length of the vehicle, and, at the discretion of the examiner, will vary within the range: plus 3 ft minus 6 ft. The precise length of the bay will not be disclosed to the candidate before completing the exercise.

● indicates 18" marker cone.
● indicates 18" marker cone with 5 ft coloured pole.
X = where the driver will position his vehicle prior to the exercise.

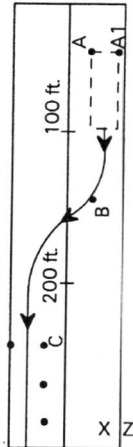

Braking Exercise

You will be allowed a distance of about 200 ft in which to attain a speed of about 20 miles an hour; at the end of this distance a marker will show you where you should begin to apply your brakes. The examiner will ride in the vehicle and will judge your ability to bring the vehicle to rest from about 20 miles an hour as quickly as possible, with safety and under complete control.

In good conditions, a well maintained goods vehicle in the hands of a competent driver should stop in the following distances from 20 mph:

Up to 4 tons unladen weight:	20 ft.
Up to 6 tons unladen weight:	25 ft.
Up to 10 tons unladen weight:	35 ft.

Make sure that the vehicle you bring for the test has good brakes. If you cannot stop within a reasonable distance when carrying out this exercise the examiner may decide to stop the test there and then in the interests of public safety; in which case you will lose your test fee.

Remember that if you apply excessive pressure on the brake pedal in an empty vehicle you may lock the wheels. This may increase your stopping distance and with an articulated vehicle may cause it to 'jack-knife'.

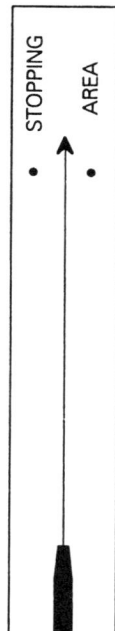

ON THE ROAD

When the candidate has completed the manoeuvres and exercises, the driving examiner will direct him out of the test centre and onto the main road. The candidate will be told to follow the main road at all times, unless told to do otherwise. In order to achieve success under test conditions it is essential that the driver concentrates on the job in hand and is not put off by the examiner's presence.

During the driving test the examiner will quietly assess the driver's ability on all types of roads and traffic conditions. Particular attention will be paid to the driver having complete control of the vehicle throughout the test, and to his ability to display courtesy and consideration to other road users at all times. It is essential to drive the vehicle safely and progressively, while adhering to all the rules of the road.

The driver must be able to adjust his driving to suit the different road conditions and also the different weather conditions. During the course of the test the examiner will ask the driver to perform various road exercises and procedures such as: gear changing, stopping and moving off and hill starts. The driver must display confidence while driving, particularly when negotiating right turns, left turns, roundabouts and traffic lights, etc. You do not have to be a good driver to pass the driving test, but you do have to be a *safe* driver.

In order to help assess the driver's ability, the driving examiner will ask the examinee to perform various 'on the road' manoeuvres, including:

- The correct procedure for stopping a large goods vehicle by the side of the road.
- Moving the vehicle safely from a

stationary position, on the level; down
hill; up hill; at an angle.

GEAR CHANGING EXERCISE

In normal driving with most large goods
vehicles, it would be unusual for a driver to
use first gear when moving off. However, in
order to demonstrate that you are able to
engage any of the low gears when necessary,
the driving examiner will require you to
perform a gear changing exercise. This is
usually conducted early on in the test. The
driving examiner will direct you to a suitable
road where this part of the test can be
conducted with a minimum of inconvenience
to other traffic. At the appropriate time you
will be asked to pull up to a convenient place
on the left, after which the driving examiner
will give you very clear instructions as to
what he requires you to do. Normally you
will be asked to move off in the lowest gear
on the vehicle and to drive for a reasonable
distance in that gear before changing up into
the next gear. This will continue until the
examiner feels that you have engaged
sufficient upward gears for the next part of
the exercise to begin. He will then ask you to
change down through each gear
progressively until the lowest gear has once
again been engaged.

It will not be necessary in this exercise to
demonstrate the use of any auxiliary
transmission systems, ie splitter box or two-
speed axles (a splitter box should not be
confused with a range change). You may use
the foot brake as necessary to slow the
vehicle down before any downward gear
change.

COMPLETION OF TEST

On completing the test successful candidates will be handed a pass certificate, which means that it is no longer necessary for the driver to be accompanied by a licence holder or to display 'L' plates.

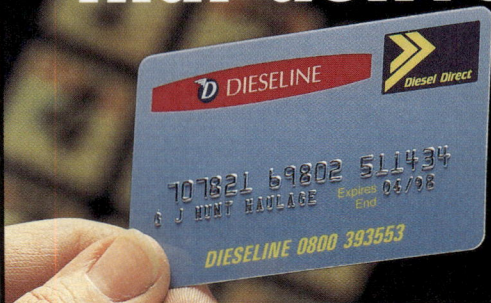

3
National Vocational Qualifications (NVQs) for Truck Drivers

Introduction To NVQs In Road Haulage
Assisting In Road Haulage And Distribution
    Operations – Levels 1–2

INTRODUCTION TO NVQs IN ROAD HAULAGE

With the introduction of National Vocational Qualifications (NVQs), the Road Haulage and Distribution Industry Training Council (RHDTC) have developed the standards for a range of qualifications specifically designed for people who are working, or who are considering working in, the transport industry.

The standards have been approved at NVQ Level 1 and NVQ Level 2 by the National Council for Vocational Qualifications (NCVQ) and the Scottish Vocational Education Council (SCOTVEC).

A person may choose any one of four qualifications that are available under the NVQ or SVQ vocational standards, depending on experience or area of work.

The qualification listings are:

• Assisting in road haulage and

distribution operations at Level 1
- Transporting goods by road at Level 2
- Storing goods for distribution by road at Level 2
- Organising road transport operations at Level 2.

In order to achieve a Level 1 qualification the candidate must demonstrate competence in ten areas; see below.

ASSISTING IN ROAD HAULAGE AND DISTRIBUTION OPERATIONS – LEVELS 1–2

Unit 1 *Contribution to the retention of business*
    1.1 Maintain effective relationships with customers
    1.2 Assist in resolving customer complaints

Unit 2 *Contribute to the maintenance of a secure and productive work environment*
    2.1 Contribute to the effective use of physical resources
    2.2 Contribute to the maintenance of security

Unit 3 *Liaising with callers and colleagues*
    3.1 Receive and assist caller
    3.2 Maintain business relationships with other members of staff

Unit 4 *Communicating information*
    4.1 Process incoming and outgoing business telephone calls
    4.2 Receive and relay oral and written messages
    4.3 Supply information for specific purpose
    4.4 Draft routine business communications

Unit 5 *Contribute to the collection and delivery of goods and materials*

5.1 Assist in the preparation of work area for transfer of goods

5.2 Assist in the transfer of goods and materials

Unit 6 *Dealing with requisition*

6.1 Establish requirements

6.2 Provide items as required

Unit 7 *Contribute to the safe movement of vehicle on company premises*

7.1 Assist in the preparation of vehicles for use

7.2 Operate and control a vehicle within company premises

Unit 8 *Contribute to the storage of goods*

8.1 Assist in the storage of goods

8.2 Contribute to the maintenance of storage conditions

Unit 9 *Contribute to the receipt and despatch of goods*

9.1 Assist in the receipt of goods

9.2 Assist in the despatch of goods

Unit 10 *Maintaining records and stock information*

10.1 Maintain records

10.2 Locate, abstract and present information for a specific purpose

In order to achieve a Level 2 qualification the candidate must first choose which of the following qualifications is most suitable (see below).

- Transporting goods by road
- Storing goods for distribution by road; or
- Organising road transport operations.

Note: A candidate may be entered directly into any of the Level 2 qualifications without having to complete Level 1.

In order to achieve any one of the above qualifications the candidate must demonstrate competence in four core units which are common to each qualification (see below).

- Contribute to the generation and retention of business
- Maintain a secure and productive work environment
- Develop and maintain effective working relationships
- Provide information to customers.

In addition, three extra units, relevant to the chosen qualification, must be completed. For example to achieve a full Level 2 in *Transporting goods by road*, the candidate must complete successfully all the following units:

The four core units (common to all Level 2 qualifications)

Unit 1 *Contribute to the generation and retention of business*
- 1.1 Contribute to the promotion of the business
- 1.2 Develop and maintain effective relationships with customers
- 1.3 Assist in resolving customer complaints

Unit 2 *Maintain a secure and productive environment*
- 2.1 Make effective use of physical resources
- 2.2 Maintain the security of physical resources

Unit 3 *Develop and maintain effective working relationships*
- 3.1 Develop and maintain effective working relationships with colleagues
- 3.2 Develop and maintain the trust and support of one's line managers

Unit 4 *Provide information to customers*
- 4.1 Respond to requests for information
- 4.2 Inform customers about available products

Transporting goods by road
Unit 5 *Prepare for the transport of goods and materials*
    5.1 Prepare vehicle for use
    5.2 Organise own work
Unit 6 *Transport goods and materials*
    6.1 Operate and control a laden vehicle
    6.2 Monitor and review progress
Unit 7 *Collect and deliver goods and materials*
    7.1 Prepare for the transfer of goods and materials
    7.2 Transfer goods and materials

To achieve a full Level 2 in *Storing goods for distribution by road,* the candidate must complete successfully all the following units:
The four core units (common to all Level 2 qualifications)
Plus:
Storing goods for distribution by road
Unit 8 *Receive goods into storage*
    8.1 Prepare for receipt of goods
    8.2 Receive goods
Unit 9 *Store goods within a facility*
    9.1 Prepare work area for storage
    9.2 Store goods
    9.3 Monitor and maintain storage conditions
Unit 10 *Despatch goods*
    10.1 Prepare consignment for loading
    10.2 Prepare work area for loading
    10.3 Load consignment onto vehicle
To achieve a full Level 2 in *Organising road transport operations*, the candidate must successfully complete all the following units:
The four core units (common to all level 2 qualifications)
Plus:
Organising road transport operations

Unit 11 *Contribution to the allocation of work*

    11.1 Assist in arranging staff cover
          to meet operational
          requirements
    11.2 Assist in the allocation of
          specific work activities

Unit 12 *Processing documents relating to goods and services*

    12.1 Reconcile incoming invoices
          for payment
    12.2 Prepare and despatch
          quotations, invoices and
          statements
    12.3 Process expense claims for
          payment
    12.4 Order office goods and services

Unit 13 *Storing and supplying information*

    13.1 Maintain an established filing
          system
    13.2 Supply information for a
          specific purpose

Training and assessment for qualifications in road haulage are based initially on the candidate, *'learning the job'*. This requirement can be met through numerous methods (see list below). The awarding body, workplace, or approved centre, all of whom can deliver the qualification, can endorse any of the following 'learning' options.

- Learning wholly in the workplace
- Learning partly in the workplace and also in an 'off the job' situation
- Accreditation of prior learning (APL).

With any of the three methods of learning, the candidates will need to prove that they are competent. This is achieved by the candidate producing a portfolio of evidence that relates to the skills that have been taught.

    In order to validate that the learning has been taught to the national standards, a qualified independent assessor judges each task identified on the standards. By using *diverse evidence* from the candidate's portfolio, and questioning the candidate about the standard and range of tasks, the assessor can gauge how sound the applicant's learning is.

Another method of ensuring competence is that of *observed assessment*. This involves the candidate carrying out a task whilst being observed by the assessor. On successful completion of all the units identified in the standards, the assessor will sign to the effect that the candidate is competent.

The next stage in the assessment procedure is internal/external verification. Once completed, the candidate's portfolio is then scrutinised by a qualified programme verifier, who ensures that all the performance criteria have been met, before countersigning the evidence portfolio accordingly.

On completion of the whole assessment process and the production of the candidate's portfolio of evidence, the relevant qualification certificate is applied for.

In order to pursue these qualifications, applicants (or their company) must apply to an *approved training organisation*. These centres must demonstrate that they can deliver competently the required training programme, along with the resource knowledge and expertise to implement a valid and reliable assessment system. In addition, an efficient administrative system which is able to maintain the standards set by the awarding body must be in place.

The approved training organisation must employ qualified programme assessors and internal verifiers trained to the relevant national standards. The Training and Development Lead Body (TDLB) is able to provide the standard of training required, due to its understanding of road haulage and distribution.

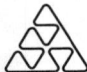

4
National Vocational Qualifications (NVQs) for Distribution and Warehousing Operations Level 2

NVQ In Distribution And Warehousing
Operations
Structure Of The Qualification
Principles Of NVQs

In order to compete effectively in the road
transport and distribution industry, many
companies are recognising the benefits of
having a properly trained and qualified
workforce.

The National Vocational Qualification
(NVQ) is an ideal vehicle for the industry to
train staff, who can then obtain a recognised
industry qualification at a level compatible
to the job that they are doing.

NVQ IN DISTRIBUTION AND WAREHOUSING OPERATIONS

NVQs are qualifications based on national standards set by industry. These standards define the skills and knowledge required for a particular area of work. Once you have met these standards, you are then competent and can get your NVQ certificate. NVQs are work-based qualifications and most of the evidence required will come from your performance at work. The standards you are working towards have been developed by the Distributive Occupational Standards Council (DOSC). City & Guilds and DOSC are the awarding body for your qualification.

NVQs are available at five different levels. The highest is Level 5. All NVQs are structured in the same way and are made up of units. A unit represents an aspect of your job.

STRUCTURE OF THE QUALIFICATION

To gain the NVQ you must complete a total of seven units:
– all three units in Group 1
– two units from Group 2
– two units from Group 3 (which must not include the units you achieved in Group 2).

Some of the units in this NVQ are common to other NVQs, such as Retail Operations Level 2. If you have completed any of the units as part of another NVQ you do not have to repeat them.

Unit numbers
The first number is the unit number allocated by DOSC. The number in brackets is the City & Guilds number which your centre will use when applying for your certificate.

Core of Mandatory units – Group 1

You must complete all three units.

Unit 1 (001)		**Contribute to the maintenance of health and safety in the workplace**
Outcome	1.1	Contribute to maintaining a healthy and safe workplace
	1.2	Implement procedures to deal with risks to health and safety
	1.3	Maintain the cleanliness of the working environment
	1.4	Manually lift and handle goods safely
Unit 2 (002)		**Contribute to the security of the workplace**
Outcome	2.1	Maintain the security of people, stock and premises
	2.2	Implement procedures to deal with risks to security
Unit 3 (003)		**Contribute to effective working relationships**
Outcome	3.1	Maintain own performance in achieving quality standards
	3.2	Work with colleagues to optimise productivity

Core or Mandatory units – Group 2

You must complete any two units from Group 2

Unit 4 (004)		**Receive and verify goods and materials entering storage**
Outcome	4.1	Prepare for the receipt of goods and materials
	4.2	Accept goods for storage
Unit 5 (005)		**Place goods and materials in storage**
Outcome	5.1	Confirm the requirements for storing goods and materials
	5.2	Place goods and materials in designated locations
Unit 6 (006)		**Assemble bulk orders for distribution to customers**
Outcome	6.1	Confirm requirements for orders
	6.2	Identify goods and materials and prepare orders to meet requirements

Unit 7 (007)	**Select stock items and assemble orders for delivery to individual customers**
Outcome 7.1	Select and assemble orders to meet customers requirements
7.2	Pack and despatch orders to customers
Unit 8 (008)	**Meet customers' needs for information and advice**
Outcome 8.1	Provide information and advice to meet customers' needs
8.2	Enable customers to resolve complaints
Unit 9 (009)	**Process customers' orders for goods**
Outcome 9.1	Confirm the availability of goods to meet customer requirements
9.2	Record and process customers' orders

Optional units – Group 3

You must complete two units from this group. These must be two different units from those completed in Group 2.

Unit 5 (005)	**Place goods and materials in storage**
Outcome 5.1	Confirm the requirements for storing goods and materials
5.2	Place goods and materials in designated locations
Unit 6 (006)	**Assemble bulk orders for distribution to customers**
Outcome 6.1	Confirm requirements for orders
6.2	Identify goods and materials and prepare orders to meet requirements
Unit 7 (007)	**Select stock items and assemble orders for delivery to individual customers**
Outcome 7.1	Select and assemble orders to meet customers' requirements
7.2	Pack and despatch orders to customers
Unit 8 (008)	**Meet customers' needs for information and advice**
Outcome 8.1	Provide information and advice to meet customers' needs
8.2	Enable customers to resolve complaints
Unit 9 (009)	**Process customers' orders for goods**
Outcome 9.1	Confirm the availability of goods to meet customer requirements
9.2	Record and process customers' orders

Unit 10 (010) **Identify and support improvements to customer service and business operations**

Outcome 10.1 Identify and recommend improvements to customer service and business operations

10.2 Help to implement improvements to customer service and business operations

Unit 11 (011) **Provide goods from stock on request**

Outcome 11.1 Select goods from stock to meet customer requirements

11.2 Process the sale of goods

Unit 12 (012) **Help customers to choose between products**

Outcome 12.1 Clarify and confirm customers' requirements for specified products

12.2 Explain the features and benefits of products likely to meet customers' requirements

12.3 Clarify and confirm custom

Unit 13 (013) **Maintain hygiene standards in handling and storing food products**

Outcome 13.1 Establish and maintain personal hygiene

13.2 Maintain the storage of good in hygienic conditions

13.3 Maintain workplace hygiene

Unit 14 (014) **Operate and maintain the effectiveness of equipment for handling and moving goods**

Outcome 14.1 Operate handling equipment to specification

14.2 Maintain records of handling equipment

Unit 15 (015) **Despatch goods and materials**

Outcome 15.1 Prepare work areas for loading

15.2 Load consignments

Unit 16 (016) **Maintain the quality of goods and materials in storage**

Outcome 16.1 Monitor and maintain storage conditions

16.2 Monitor goods and materials in storage

PRINCIPLES OF NVQs

The principles of NVQ qualifications are based on what a person can do, what that person knows about the industry, and the specific job that is being or about to be done.

As there are no formal examinations involved, a person must be able to demonstrate competence in a whole range of units and elements. This is achieved by the candidate producing a portfolio or evidence.

The candidate must then state *how* the performance criteria have been met (I will discuss this later) and provide the evidence to prove it.

Each qualification is made up of a series of separate units. Each unit identifies exactly what the candidate must be able to do and to what standard. Training and assessment for qualifications in wholesale, warehousing and stores are based on the candidate first '*learning the job*'.

The method of learning can vary according to the candidate, the awarding body, the workplace, or the approved centre who would deliver the qualification. Learning may follow any of the approved methods:

- learning wholly in the workplace
- learning partly in the workplace and partly in an 'off the job' situation
- accreditation of prior learning (APL).

With any of the three methods of learning the candidate will need to prove competence by producing a portfolio of evidence that relates to the skills that have been taught.

In order to validate that the learning has been taught to the national standards, a qualified independent assessor assesses each task identified on the standards. This is achieved by the assessor *using diverse evidence* from the candidate's portfolio and questioning the candidate against the standard and the whole range of the tasks.

Another method of ensuring competence is that of *observed assessment*. This is achieved by the candidate doing the task and being observed by the assessor. On successful completion of all the units identified in the standards, the assessor will sign to the effect that the candidate is competent.

The next stage in the assessment procedure is internal/external verification. A completed portfolio is scrutinised by a qualified programme verifier, who ensures that all the performance criteria have been met before countersigning the evidence portfolio accordingly.

On completion of the whole assessment process and the production of the candidate's portfolio of evidence, the relevant qualification certificate will be applied for.

In order to pursue these qualifications, applications (or their company) must apply to an *approved training organisation*, which has demonstrated competence for the delivery of this training programme and has further demonstrated that it has the resources, knowledge and expertise to implement a valid and reliable assessment system in addition to an administration capable of maintaining the standards as required by the awarding body.

Approved training organisations will employ qualified programme assessors and internal verifiers trained to the relevant national standards of the Training and Development Lead Body (TDLB). They must be occupationally competent in wholesaling, warehousing and stores.

For more details of the NVQ Qualification contact the local Training and Enterprise Council.

5
Carriage of Dangerous Goods

Training Requirements

The law relating to the carriage of dangerous goods is contained in the following regulations:

- ADR – European Regulations
- CDG Road – Carriage of dangerous goods by road regulations 1996: covers carriage in tankers, packages and bulk
- DTR – Carriage of dangerous goods by road (driver training) regulations 1996
- CDG/CPL – Carriage of dangerous goods (classification packaging and labelling) and use of transportable pressure receptables regulations 1996
- RAM Road – Carriage of radioactive materials (Great Britain) regulations 1996
- CER – Carriage of explosives by road regulations 1996
- ACL – Approved carriage list: information approved for the carriage of dangerous goods by road and rail other than explosives and radioactive material
- ATR – Approved tank requirements: the provisions for bottom loading and vapour recovery systems of mobile containers carrying petrol

All of the above came into force on 1 September 1996.

The regulations apply to anyone involved in the transporting of dangerous goods by road including drivers, fork lift truck operators and warehousing staff.

The regulations are quite comprehensive but in broad terms it is the responsibility of the operator to ensure that:

- written information includes a transport document and a signed statement confirming that all the regulations are being complied with
- all the necessary precautions are taken when storing, loading and unloading to prevent the risk of fire, breakage or explosion
- the vehicle carrying the dangerous goods is suitable and that the orange marker plates are available and used
- the driver has been properly trained and certified in the carriage of dangerous goods.

Drivers who are involved in the transporting and distribution of dangerous goods are responsible for ensuring that:

- information relating to the dangerous goods, including the transport emergency card (TREMCARD) and transport documents, is kept in the vehicle at all times
- the orange marker plates and hazard warning panels are displayed accordingly, and that any information regarding the dangerous goods is given to the appropriate authority if required, eg police officers and fire officers
- the vehicle and load is protected against fire and explosion at all times including safe parking and safe loading and unloading

TRAINING REQUIREMENTS

Under the Road Traffic (Carriage of Dangerous Goods by road (Driver Training) Regulations 1996) Regulations, drivers are required to undergo training by attending an approved course and passing the appropriate examinations in order to obtain the ADR Vocational Training Certificate.

Training courses are based on theoretical training sessions and practical exercises. The duration of training is determined by the number of modules a driver needs to cover in accordance with the type of work being carried out. For example, a driver involved in packaged goods only, will need to pass:

(a) core module
(b) packages module
(c) at least one class or classes as required

A driver involved in road tankers and tank containers only will need to pass:

(a) core module
(b) tanker specialisation module
(c) at least one class or classes as required

A driver involved in both tanker and packaged goods will need to pass all of the above modules.

The additional modules include:

Class 1	–	Explosives (Specialisation Module)
Class 2	–	Gases
Class 3	–	Flammable liquids
Class 4.1	–	Flammable solids
Class 4.2	–	Spontaneously combustible
Class 4.3	–	Dangerous when wet
Class 5	–	Oxidising agents and organic peroxides
Class 6	–	Toxic and infectious
Class 7	–	Radioactives (Specialisation Module)
Class 8	–	Corrosives
Class 9	–	Miscellaneous

Drivers who successfully complete the training course and subsequently pass the relevant examination will be issued with the Vocational Training Certificate. The certificate is valid for five years, although one year prior to the renewal the driver will need to take a refresher course and pass the relevant examination again.

For more details regarding the carriage of dangerous goods and the training courses, see 'Dangerous Goods – Driver Training Centres' in the Appendix.

In addition to the panel shown below, a secondary hazard warning sign (diamond) must be affixed on a level plane with the panel. This sign (diamond) will not have a number at the bottom tip and must measure 200 mm × 200 mm.

The panel illustrated here is for an inflammable gas. The numbers and letters in the top column are a code which identify the type of equipment required to deal with the chemical in the event of an emergency. The numbers in the bottom column are a code which identify the chemical. The information at the bottom of the panel gives the point to be contacted in the event of an emergency. The information is designed to help police and fire services to deal with an incident in the appropriate way.

Markings for tankers and bulk loads are basically a large panel split up with the warning signs (diamonds) measuring 250 mm × 250 mm which must be on a level plane with the halved panel.

Under the CDG–Road regulations 1996 an orange reflectorised panel measuring

400 mm × 300 mm with the black border
must also be affixed to the front of the
tanker or bulk load container.

Warning Signs (Diamonds)
The signs shown below are for secondary
hazards.

Primary hazard signs have the class
number shown in the bottom tip.

Substances liable to ignite spontaneously (red/white)	Poisonous substances. (white)	Oxidizing substances (yellow)
Radioactive substances (yellow)	Compressed gases (green)	Corrosive substances (black/white)

IF A SPILLAGE OCCURS – KEEP WELL AWAY
AND INFORM THE POLICE OR THE FIRE BRIGADE

Figure 5
Signs indicating dangerous substances

6
Tachographs – Questions and Answers

Domestic Rules
Questions And Answers

Courtesy of Lucas Kienzle Instruments Ltd
Reproduced from *The Tachograph Manual*
(second edition) by David Lowe, published by
Kogan Page.
Figure 6
**An external view of an
approved-type tachograph**

There are three major tachograph manufacturers in the UK – TVI-Europe Ltd, Time Instruments Manufacturers Ltd and Lucas Kienzle Instruments Ltd. Figure 6 shows a Lucas Kienzle tachograph, but all are essentially the same.

This chapter explains tachograph functions in a straightforward way. You should not, of course, regard the information as a legal authority.

Tachographs *must* be checked and calibrated at a Government-approved calibration station.

1. *Opening key*
   The key for opening and closing the tachograph is usually kept in its lock. It is important that the key is available to open the tachograph for inspection at any time during the driver's working day.

2. *Driver mode selector*
   Use this control to record the type of your activity (see Figure 7).

3. *Speed warning light*
   A small control (located inside the tachograph) can be adjusted to bring on the speed warning light at almost any speed the drive chooses. Its purpose is

Active work other than driving (not in use in the UK)

Duty time

Actual driving time spent behind the wheel

Any periods of rest

Figure 7
Standard tachograph modes

to advise the driver that the pre-set
speedometer reading has been reached.

4. *Total odometer*
   The total odometer records the total
   distance travelled by the vehicle in
   kilometres, in the same way as a
   milometer records the total distance
   travelled in miles.

5. *Electronic clock and clock operation
   indicator*
   This is usually electronic quartz. The
   operation indicator shows at a glance
   (red and white moving stripes) whether
   or not the clock is working.

6. *Chart illuminating switch and chart
   viewing window (optional)*
   By pressing the illumination switch the
   driver can see the recordings as they are
   registered on the chart (see Figure 8).

7. *Fourth stylus operation warning light*
   An optional stylus offers an additional
   trace on the chart. For example, if an
   ancillary engine is used on a tanker, or
   for dropping the brush on a sweeper, or
   for tilt-lift operations, etc.

8. *Trip odometer*
   This can be set to record the distance
   travelled per trip in kilometres.

9. *Dual scale*
   This is a speedometer showing miles per
   hour up to 70 mph and kilometres per
   hour up to 125 kph.

10. *The chart*
    You must complete the centre field of the

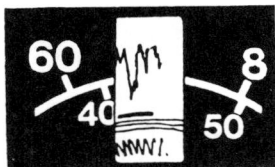

Figure 8
Chart viewing window

chart *before* installing it in the tachograph, and before commencing a journey. These are the details that you should enter:

- Full name
- Registration number of vehicle
- Date at start of work
- Initial odometer reading
- Place of departure

At the end of the working day you must remove the chart and complete the centre field by adding:

- Total odometer reading
- Date at end of day
- Finishing location
- Total distance covered

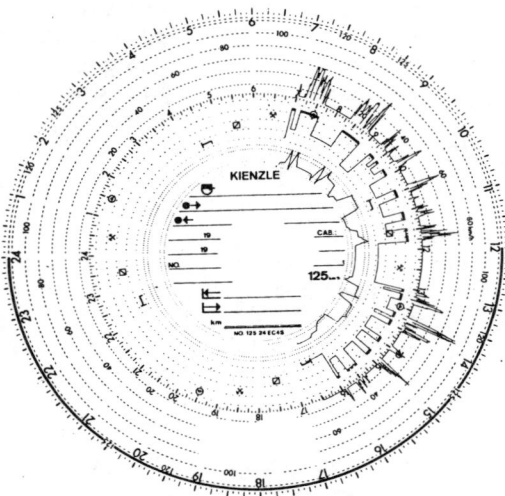

Figure 9
Chart analysis

Reproduced from *The Tachograph Manual* (second edition) by David Lowe, published by Kogan Page.

Figure 9 shows the centre field where this information should be recorded, and the information which is scribed on to the chart by the tachograph's three styluses.

QUESTIONS AND ANSWERS

Drivers may find the following typical questions, and the appropriate answers, helpful.

Does the driver have a key which opens and closes the head?

Yes, he must. He has to be able to open the case and insert and remove charts. He may open the case to check on his hours. He may also be asked by a police officer or traffic examiner to open the case so that they can check the chart.

Can recordings be made without the driver's knowledge?

No. It is one of the requirements for EU type approval that it should be possible to tell, without even opening the tachograph case, whether recordings are being made.

Can the records be falsified?

It is virtually impossible to falsify tachograph records in a way which will deceive a trained examiner. Do not forget that falsification of tachograph records is an offence, which would make you liable to prosecution.

Can I be prosecuted for speeding on the basis of what the tachograph chart shows?

No. A tachograph record would not be enough on its own to obtain a conviction for speeding. It could be used in corroboration with other evidence – a police patrol or radar check, for example.

To bring about a prosecution for speeding, evidence must be produced showing:

- the speed at which the vehicle was travelling (which the chart confirms)
- the time at which the vehicle was travelling at the alleged speed (which the chart also confirms)
- the precise location at which the vehicle was travelling at the alleged speed (which the chart *does not* show.)

It is because of this missing piece of vital evidence that drivers will not be prosecuted for speeding on the basis of tachograph chart readings alone.

What entries must the driver make on the chart?

1. Fill in the required information on the centre of the chart. This is important – you must make these entries or you will be breaking the law.
2. After ensuring that the clock is set to the right time, insert the chart into the instrument, checking am/pm are true to time.

3. During the day, turn the mode switch to show what you are doing – driving, other work, and breaks from work or rest period.
4. The tachograph itself will automatically record the speed and distance travelled. At the end of the day, take out the chart and write in the place of finishing work and the odometer reading.
5. Keep the chart for at least two days, then hand it back to your employer within seven days of completing it. *The entries on the centre field are very important – do not forget them.*

What if I am working away from the vehicle?

● Keep the chart with you and make manual entries on it, *or*
● If no one else is going to use the vehicle, and you are going to be engaged in only one kind of activity while you are away from the vehicle, you can leave the chart in the tachograph, with the mode key set to the appropriate activity, and the tachograph will record the time automatically.

What if I change vehicles?

If the vehicle has a tachograph of the same type, you can carry on using your present tachograph chart; if not, use the appropriate chart and make manual entries. Note the time of the vehicle change, the registration number of the new vehicle, and the odometer readings on both vehicles in the space provided on the chart.

How are daily and weekly rest periods shown on the charts?

The law requires that information on rest periods should be entered on the chart, but it does not say how this should be done. In some cases, it may be possible to leave the chart in the tachograph overnight, with the mode key set at 'rest'. The tachograph itself will then record the daily rest period on the chart.

Start a new chart when you come on duty in the morning. If this is not possible, you should maintain a manual record or make a note on the chart at the beginning and end of the rest period – eg 'start daily rest 06.15' and 'end daily rest 20.30'. Use the 24-hour clock.

Tachograph charts should *never* be left in a tachograph for more than 24 hours. For weekly rest periods, or longer periods away from work, make a manual note on the tachograph chart at the end of the last day's work and at the beginning of your first day back.

What if there are two drivers in the cab?

If there are two drivers in the cab, and the vehicle is fitted with a two-man tachograph, then both charts (driver and mate) should be inserted into the tachograph. If not, manual entries must be made on a chart by the second man.

In the case of a two-man tachograph, each driver inserts his own chart. On the driver's chart, the tachograph records speed and distance as well as time and work modes. On the other, only time and work modes are shown.

What should the driver do if his tachograph chart is seized by enforcement staff?

The law does not lay down any procedure for this, but it is suggested that the driver should ask for a receipt showing the date, time and name of the officer who took the chart. He should then give this receipt to his employer to keep in place of the chart itself. A new chart must be put into the tachograph before the driver continues his journey.

How and where do I make manual entries?

The law says only that the entries must be legible and that you must not use a dirty chart.

How you do it depends on which type of chart you are using. Some manufacturers recommend making entries on the back of the chart and provide a specially printed area for this. Others recommend making entries on the front of the chart, round the outside edge.

Whatever method you use, take care not to damage existing recordings on the chart. Since space is limited, it maybe helpful to use abbreviations – D for driving, OW for other work, etc.

How do I know that the equipment is working?

Each manufacturer provides some means of checking that the tachograph is making recordings. In some types, there is a window through which you can see the chart, with a light to illuminate it.

What if the equipment breaks down?

Make manual entries on the chart. The tachograph must be repaired either on return to base or within seven days. But if it is simply not possible to get the tachograph repaired at once, you can go on using the vehicle in the meantime – the law provides a defence against conviction to cover this. You must keep a manual record until the tachograph is repaired.

What if the chart gets damaged or dirty?

Do not use it. If the chart gets damaged or dirty when you have already made recordings on it, start a new chart and attach the old one to it at the end of the day.

7
Drivers' Hours and Records

General
Penalties For Offences
EU, International And National Rules
Double Manning Of Vehicles
Transport By Ferryboat Or Train
Emergencies
Exemptions
Domestic Rules

GENERAL

You must observe rules laid down in law on
how much driving and other work you can
do and how much rest you must take
between working periods. The law also
makes your employer responsible for
ensuring that you observe these rules. Very
few goods vehicle drivers are completely
exempt from drivers' hours regulations.
    These are:
- drivers of vehicles used by the armed
  forces, the police and fire brigades (but
  not other crown vehicles)
- drivers whose driving is done completely
  off the public road system
- drivers who are not driving in connection
  with a job or in some way to earn a
  living.

There are different sets of rules for different
types of work and sizes of vehicles, based on
domestic legislation contained in the

Transport Act 1968 and EU legislation. The sets of rules are as follows:

EU National and International rules	Apply to drivers of most vehicles of over 3.5 metric tonnes permissible maximum weight.
AETR rules	Apply to drivers of vehicles on journeys to, or through, some European countries outside the EU (very similar to EU).
Domestic rules	Apply to drivers of most vehicles up to 3.5 tonnes permissible maximum weight and drivers whose activities are exempt from the EU Rules.
Mixed driving	Apply to drivers who carry out operations covered by more than one of the above groups.

Goods vehicles include virtually all vehicles constructed or adapted to carry or haul goods. If you also work with passenger vehicles, then in general the law treats you as a goods vehicle driver unless you are spending more than 50 per cent of your time working with passenger vehicles which are subject to drivers' hours law; and work done with such passenger vehicles normally counts towards your total daily and weekly allowance.

PENALTIES FOR OFFENCES

There is a maximum fine of £1,000 for offences against the rules on drivers' hours and records. This applies to drivers and to anyone whose orders the driver was following. These offences are not endorsable on driving licences, but convictions may be taken into account by licensing authorities in deciding whether to grant or review an operator's or LGV driver's licence.

EU INTERNATIONAL AND NATIONAL RULES

Daily driving limit

You must not drive for more than nine hours in a day, but this may be extended to ten hours twice a week. The daily driving limit means the period spent at the wheel between any two daily rest periods or between a daily rest period and weekly rest period. Driving off the public roads does not count as driving time.

Continuous driving limits and breaks

After four and a half hours' driving, whether non-stop or in several shorter periods, you must take a break of at least 45 minutes, unless you begin a daily or weekly rest period. This break may be replaced by breaks of at least 15 minutes each totalling 45 minutes spread out over the 4½ hour driving or immediately after it. During a break you cannot do any work but waiting time and time not devoted to driving spent in a vehicle in motion, a ferry or a train, shall not be regarded as 'other work'.

Legal 1

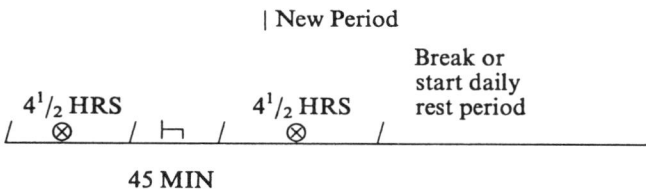

Legal 2

| New period | New period

1 HR $4^1/_2$ HRS $3^1/_2$ HRS

/ ⊗ / ⊢ / ⊗ / ⊢ / ⊗ /

45 MIN   45 MIN

Legal 3

| New Period

Break or
start daily
rest period

30 MIN   4 HRS $4^1/_2$ HRS

/ ⊗ / ⊢ / ⊗ / ⊢ / ⊗ /

30 MIN   15 MIN

Illegal

After 30 mins driving
| New period | ILLEGAL

Break or
start daily
rest period

30 MIN   4 HRS $4^1/_2$ HRS

/ ⊗ / ⊢ / ⊗ / ⊗ /

45 MIN

Weekly driving restriction

There is no weekly driving limit as such, but a weekly rest period must be taken after not more than six daily driving periods. The weekly rest period may be postponed until the end of the sixth day if the total driving time over the six days does not exceed the maximum corresponding driving periods.

Fortnightly driving limit

The total period of driving in any one fortnight shall not exceed 90 hours. This means that if you drive for a total of 48 hours this week, next week you would only be able to drive for 42 hours.

Daily rest

You must have a minimum rest of 11 consecutive hours in any 24 period but this may be reduced to 9 hours not more than 3

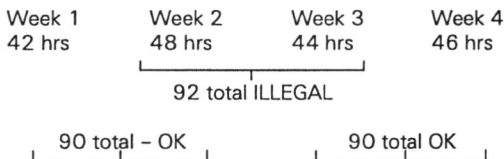

Week 1	Week 2	Week 3	Week 4
42 hrs	48 hrs	44 hrs	46 hrs

92 total ILLEGAL

90 total – OK   90 total OK

Figure 10

times a week, as long as the reduction is made up by an equivalent rest before the end of the following week. Alternatively, 12 hours daily rest may be taken in two or three periods, one of which must be at least 8 hours, and all of which must be at least one hour. The daily rest period may be taken in a vehicle, as long as it is fitted with a bunk and is stationary.

Minimum weekly rest

You must have a minimum weekly rest of 45 consecutive hours; this can be reduced to 36 hours if it is taken at your home or base or 24 hours if taken elsewhere. Each reduction shall be compensated by an equivalent rest taken 'en bloc' before the end of the third week following the week concerned. A weekly rest period which begins in one week and continues into the following week may be attached to either of these weeks.

Compensation for reductions in rest periods

Any rest taken as compensation for the reduction of the daily and/or weekly rest periods must be attached to another rest of at least eight hours. You may choose to take it at either the vehicle's parking place or your base. When taking compensation for daily and/or weekly rest, you should make your tachograph chart for that day accordingly, eg 'Compensation for DR/WR for (date) (number of hours/minutes)'.

DOUBLE MANNING OF VEHICLES

The same rules apply to vehicles continuously manned by two drivers except that during each period of 30 hours each driver shall have a rest period of not less than 8 consecutive hours. In a double manned vehicle, one driver may take a break (but *not* a daily rest period) on the moving vehicle while the other is driving.

TRANSPORT BY FERRYBOAT OR TRAIN

When the daily rest period includes time spent on a ferryboat or train it may be interrupted once only, so that part may be taken on board the ferryboat or train and part elsewhere. The period of interruption must be as short as possible and must not exceed one hour before getting on or off including going through Customs. During both portions of the rest period you should have access to a bunk or couchette and the daily rest period must be increased by two hours. Time spent on a ferryboat or train which is not treated as daily rest may be regarded as a break.

EMERGENCIES

Provided that road safety is not jeopardised, and to enable you to reach a suitable stopping place, you may depart from the driving limits and rest requirements to the extent necessary to ensure the safety of persons, of the vehicle or of its load. A note of the occurrence and the reason why the requirements were disregarded must be made on the back of the tachograph chart.

EXEMPTIONS

Several categories of vehicle are exempt from EU Rules either for National or both National and International operations. These categories include:

- highway maintenance
- refuse collection and disposal
- breakdown vehicles
- road test
- milk collection from farms
- local agricultural work
- animal waste not intended for human consumption
- driver training
- some livestock movements.

The full list mentions a large number of specialised operations. Your employer will confirm if exemptions apply to you; if in doubt please ask your employer for further details.

DOMESTIC RULES

These rules apply to drivers of goods vehicles which are exempt from EU rules.

Daily driving limit

You must not drive for more than 10 hours in a day. The daily driving limit applies to time spent at the wheel, actually driving. Off-road driving for the purpose of agriculture, quarrying, forestry, building work or civil engineering counts as duty rather than driving time.

Daily duty limit

You must not be on duty for more than 11 hours on any working day. You are exempt from the daily duty limit on any working day when you do not drive. If you do not drive for more than four hours on every day of the

week you are exempt from the daily duty
limit.

Mixed driving

Where you use a vehicle which is subject to
the EU rules during a day or week in which
you also drive a vehicle subject to domestic
rules, you may either observe the EU hours
rules all the time, or a combination of both
rules as long as the EU limits are not
exceeded when driving vehicles on EU work.
The following points must be considered:

- the time spent driving or on duty under
  EU rules cannot count as an off duty
  period under domestic rules
- the time spent driving or on duty under
  domestic rules cannot count as a break or
  rest period under EU rules.
- any EU driving in a week means that the
  driver must take EU daily and weekly
  rest.

Emergencies

The domestic rules are relaxed for events
needing immediate action to avoid danger to
life or health, serious interruption of
essential public services (gas, water,
electricity or drainage), or of
telecommunications and postal services, or in
the use of road, railways, ports, airports, or
serious damage to property. In these cases
the driving and duty limits are suspended for
the duration of the emergency.

Please note:
This section is a guide to drivers' hours.
Should you require any clarification or other
help, contact your transport manager or
perhaps the RHA who so kindly offered
their assistance in completing this chapter.

8
Road Transport
Regulations

Vehicle Prohibitions
Height Limits
Lights And Markers
Plating
Other Fittings
Before You Go – A Checklist
Definition Of Terms
Wide And Long Loads

VEHICLE PROHIBITIONS

The system of vehicle prohibition notices is covered by the Road Traffic Act.

The system lends itself to the protection of others against vehicles with serious and not so serious defects.

A series of vehicle prohibition notices may be issued to a driver and/or the vehicle operator demanding immediate action.

Serious consequences can result from the imposition of a PG Notice, and should such a notice be issued against the vehicle the driver must inform the operator immediately.

Prohibition notices are as follows:

GV3 direction to proceed to a place for inspection

Should a vehicle examiner wish to inspect a vehicle, the site may not always be suitable. In such circumstances, the examiner will issue a GV3 direction to the driver to take the vehicle to a specified place where the vehicle may be inspected more thoroughly, not more than five miles from the site of the first preliminary inspection.

PG9 immediate or delayed prohibition

A vehicle examiner or an authorised police officer may issue an immediate prohibition notice if it is considered that the vehicle is unfit for service.

A delayed prohibition notice will take effect from a specified date not less than 10 days from the original inspection.

Note: The LA may take further action in respect of an operator's licence as a result of prohibitions issued against vehicles.

PG9 A, B or C – variation in terms of a combined refusal

This is used where a refusal to remove a prohibition is made and/or there has been a variation of time of a prohibition.

This notice should be carried on the vehicle and must be given to an examiner when requesting that a prohibition notice be removed.

PG9B permitting movement of a vehicle

Where a vehicle has been issued with a PG9 immediate prohibition notice it may be necessary to move the vehicle to a place for repair and in such circumstances, a PG9B will be issued permitting movement of the vehicle providing the following conditions are met:

- the vehicle is unladen
- the vehicle is not towing a trailer
- the vehicle is towed by rigid tow bar or suspended tow
- the vehicle proceeds directly to the nearest specified place for repair.

PG9D – defect continuation sheet

This is a continuation sheet that would list additional vehicle defects that could not fit onto the PG9 Notice.

PG10 – removal of a prohibition

Where a vehicle has been presented for inspection after all identified defects have corrected and the vehicle has been found to be fit for service, a PG10 will be issued that will remove the prohibition.

Note: Driving a dangerously defective vehicle on the road will render the driver liable to prosecution and may also render the vehicle's insurance inoperative.

HEIGHT LIMITS

- There is *no height limit* for a large goods vehicle
- Vehicle and loads in excess of 3.7 metres (12 ft) overall height must display in the driver's cab a notice giving the total height in feet and inches
- The minimum height of an unmarked bridge is 5.03 metres (16ft 6in). All bridges below this height must be clearly marked
- The minimum height of an overhead electric cable is 5.8 metres (19ft).

LIGHTS AND MARKERS

Lights

All lights shown to the rear of a vehicle must be *red*, except those used:

- to indicate reversing
- to illuminate the interior of a vehicle
- to illuminate a rear view number plate
- to illuminate a PCV destination board.

All lights shown to the front of a vehicle must emit a *white* or *amber* light, with special provisions for service vehicles such as police, fire, ambulance and other vehicles where the use of a blue light is permitted. All motor vehicles require:

- two headlamps
- two side lamps
- two rear lamps
- two red reflectors.

Stop lights
- Vehicles first used after January 1971
  must have two red stop lamps at least 600
  millimetres apart.

Hazard warning lights
A driver may use the hazard warning lights
when the vehicle is stationary, ie loading or
unloading, to warn other road users that the
vehicle is causing a temporary obstruction.
They may also be used when the vehicle is
moving to warn traffic behind of a build-up
of stationary vehicles in front and that your
vehicle is likely to be slowing right down to a
stop. This usage occurs most often on
motorways or other roads carrying fast-
moving traffic.

Daytime lights
Drivers are required by law to use daytime
lights when visibility is seriously reduced by
fog, heavy rain, spray, snow, or other
adverse conditions. It is *your* responsibility
to decide when lights are needed. The rule is:
see and be seen.

Fog lamps
Rear: Most vehicles should be fitted with one
or two rear fog lamps.
Front:
- Fog lamps fitted to the front of a vehicle
  must emit a white or amber light
- They must work as a pair through one
  independent switch
- Lamps fitted less than 609.6 mm (2 ft)
  from the ground should only be used in
  conditions of fog or falling snow.

Rear fog lamps
Drivers should not use the high intensity rear

fog lamps unless visibility is seriously reduced, ie where visibility is less than 100 metres. Do not use them simply because it is dark, or raining or misty.

Lights on projected loads
Side projection: Where a load overhangs a side of the vehicle, front position lights must be carried within 400 mm of the edge of the load.

Rear projection: Where a load projects more than 1 m beyond the vehicle's rear position lights an additional rear lamp must be carried within 1 m of the end of the load. Where rear projections obscure the vehicle's rear lamps and reflectors, additional lamps and reflectors must be fitted to the load.

Long loads: Vehicles (and combinations) in excess of 18.3 m in length (including load) must carry side marker lamps on each side of the vehicle or load. One lamp on each side must be within 9.15 m of the front of the vehicle or load. Another, on each side of the vehicle, must be within 3.05 m.

Additional lamps must be carried to ensure that there are no gaps of more than 3.05 m.

Markers
Rear reflective markers (Figure 11) must be fitted to goods vehicles
- first used before 1 August 1982 over 3050 kg unladen weight, or
- with a maximum gross weight of more than 7500 kg
  and trailers
- manufactured before 1 August 1982 with an unladen weight of 1020 kg or more, or
- with a maximum gross weight of 3500 kg or more.

Specific markers are as follows:

Diagram 1

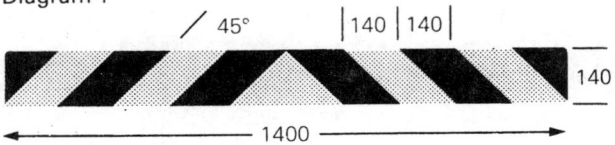

45° |140|140|

140

1400

Diagram 2

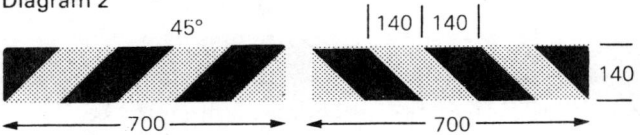

45° |140|140|

140

700   700

Diagram 3

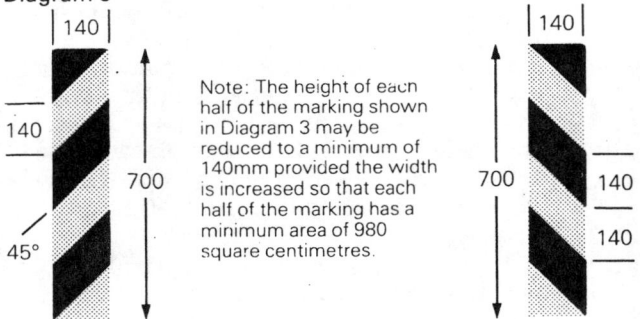

|140|

140

700

45°

|140|

700

140

140

Note: The height of each half of the marking shown in Diagram 3 may be reduced to a minimum of 140mm provided the width is increased so that each half of the marking has a minimum area of 980 square centimetres.

Diagram 4

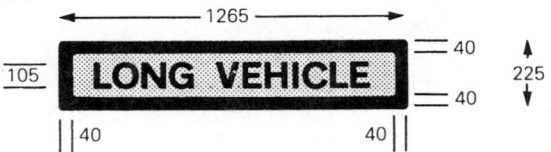

1265

105

LONG VEHICLE

40

40

225

||40| 40||

Diagram 5

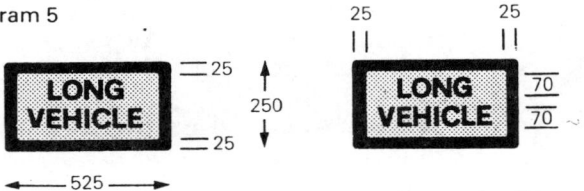

LONG VEHICLE

25

25

250

525

25   25
|| ||

LONG VEHICLE

70

70

Dimensions are shown in millimetres

Figure 11

Vehicles not exceeding 13 m in length, or trailers which form part of combinations of vehicles not exceeding 11 m – markers shown in diagrams 1, 2 or 3

- Trailers which form part of combinations of vehicles exceeding 11 m, but not exceeding 13 m – markers shown in diagrams 1, 2, 3, 4 or 5
- Vehicles exceeding 13 m or trailers forming part of combinations exceeding 13 m – marker shown in diagrams 4 or 5.

Markers *must be clean* and clearly visible from the rear at all times. When sheeting and roping, take special care not to obscure the plates. Where rear reflective markers are obscured by the load they can be fitted to the rear of the load rather than to the vehicle.

PLATING

A manufacturer's plate must be fitted to all vehicles (with certain exceptions) and must display the following information:

- Manufacturer's name
- Vehicle type
- Engine type and power
- Chassis or serial number
- Number of axles
- Maximum axle weight for each axle
- Maximum gross weight
- Maximum train weight
- Maximum weight in Great Britain for each axle
- Maximum gross weight in Great Britain.

Trailers for which goods vehicle test certificates have been issued under the plating and testing regulations must carry a current test date disc in a conspicuous position on the nearside of the trailer. Trailer plates fitted to trailers must display the following information:

- Manufacturer's name
- Chassis or serial number
- Number of axles
- Maximum weight for each axle
- Maximum load imposed on drawing vehicle
- Maximum gross weight
- Maximum gross weight in Great Britain for each axle
- Maximum gross weight in Great Britain
- Year of manufacture.

OTHER FITTINGS

Noise
The law requires all vehicles to be fitted with a silencer capable of reducing the noise from the engine to these levels:
- 92 decibels on vehicles weighing more than 3,560 kg gross plated weight
- 88 decibels on vehicles weighing less than 3,560 kg gross plated weight
- 87 decibels on passenger vehicles, including motor cars.

Vehicles travelling to a testing point or maintenance depot are exempt.

Speedometer
Vehicles must have either a speedometer or a tachograph complying with EU law.

Tachographs
Tachographs must be calibrated to ensure accurate operation. A plaque confirms the date of last calibration and 2-yearly 'check'.

Excessive smoke and vapours
It is an offence for any vehicle to emit excessive amounts of smoke or visible vapours, gases or other substances which may cause danger or injury to others. Excess-fuel devices must *not* be used on diesel engine vehicles while in motion.

Tow rope

Vehicles towing by means of a two rope or chain must not be more than 4.5 m apart. Where the distance exceeds 1.5 m the tow rope or chain must be made clearly visible on both sides. There is no stated length for a rigid tow bar.

Ancillary equipment

Ancillary equipment must *not* be used when the vehicle is in motion, except in cases of power take offs. Vehicles operating **PTO** equipment may be left unattended for short periods with the engine running *only if it can be done so safely*.

Attendants

- Attendants are *not* required on drawbar outfits, provided the combination is fitted with power brakes or over-run brakes
- Where the vehicle is towing more than one trailer an attendant *is* required
- Attendants are also required for certain long, wide or projecting loads.

Windscreen wipers and washers

- Wipers must be efficient and must afford an adequate view of the road to the front, the front-offside, and the front-nearside
- Washers must be capable of keeping a windscreen clean. It is an offence to have an empty windscreen washer bottle.

Spray suppression equipment

Spray suppression equipment must conform to British Standard Specifications and apply to:

- Vehicles exceeding 12 tonnes gross vehicle weight first used from April 1989
- Trailers exceeding 3.5 tonnes gross weight manufactured from May 1985
- Trailers exceeding 16 tonnes gross weight.

Doors

A driver is responsible for the opening of *all* doors on his vehicle. It is an offence to open a door in such a way as to cause damage or injury to any person.

Horns

- All vehicles must be fitted with a horn
- Horns must not be sounded while vehicles are stationary (except in emergencies)
- Horns must not be sounded at night (between 11.30 pm and 7.00 am) in a built-up area (except in emergencies)
- Vehicles first used after 1 August 1973 must not be fitted with musical or multi-tone horns.

Audible warning devices

These should not be used in a built-up area between the hours of 11.30 pm and 7.00 am. Although there is no legal requirement to have such devices fitted, many local authorities and other transport operators consider them to be essential in the interests of road safety.

Tyres

All tyres must be suitable for the vehicle, be properly inflated and free from cuts and other defects. They must have minimum tread depths of:

- 1.6 mm for vehicles up to 3.5 tonnes gross weight
- 1 mm for vehicles over 3.5 tonnes gross weight

Side guards

Side guards must be fitted to all heavy goods vehicles and trailers including semi-trailers, with certain exemptions. The local HGV testing centre can advise on which vehicles are exempt.

BEFORE YOU GO – A CHECKLIST

- Check your vehicle – lights, horn, oil, water, tyres, washers and wipers, couplings (if any), fuel, low pressure warning devices
- Check your load is *safe* and *secure* – trailer locks, roping and sheeting – and *secure all fastenings*
- Are vehicle and load fully equipped with all the necessary plates and markers?
- If the vehicle has a trailer make sure that the registration number corresponds with that of the tractive unit
- Do you know the height of your vehicle?
- Have you got all the necessary paperwork?
- Have you entered all the appropriate information on your tachograph chart, and ensured that the tachograph is working properly?
- Do not forget to ring in when empty.

DEFINITION OF TERMS

PTO   Power take-off. A device that transfers the power from the engine to operate auxiliary equipment.

Trailer   A trailer designed and constructed to carry goods being drawn by, and not part of, a towing vehicle, ie a vehicle and trailer – a wagon and drag.

Semi-trailer A trailer designed and constructed to carry goods being drawn by, and being an integral part of, the vehicle, ie an articulated vehicle.

GVW   Gross vehicle weight – the overall weight of the vehicle and its load.

| *ULW* | Unladen weight – the overall weight of the vehicle only. |
| *Kerbside weight* | The overall weight of the vehicle – unladen with a full tank of fuel and carrying any tools and equipment that are normally carried in vehicles. |

WIDE AND LONG LOADS

Where loads project beyond the vehicle the following requirements must be met.

End marker boards
Where these are required (see Figure 12 for an example):
- they must be placed at the *end* of the projection, or within 1 m (3ft) of the forward projection and 60 cm (2 ft) of the rear projection
- they must not be more than 2.5 m (8ft 6 in) from the ground
- they must interfere as little as possible with the driver's vision.

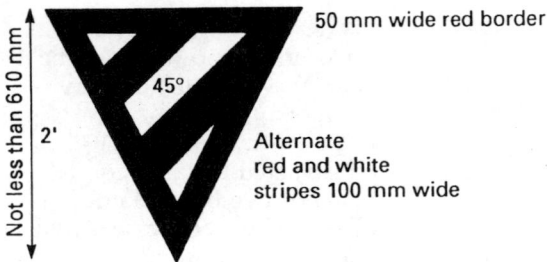

Not less than 610 mm

2'

50 mm wide red border

45°

Alternate red and white stripes 100 mm wide

Figure 12
End Marker Board

Side marker boards
Where these are required (see Figure 13):
- they must be placed within 1 m (3ft) of the end marker board on forward

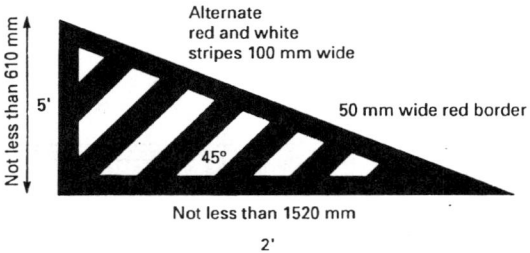

Figure 13
Side Marker Board

projections, and/or 1 m (3.3 ft) of the end
marker board on rear projections
- they must be clearly visible from *each* side
of the projecting load
- additional side marker boards should be
placed within 2.4 m (7.9 ft) of the first
marker board on forward projections,
and within 3.6 m (11.8 ft) on rear
projections.

Note: All marker boards must be clearly
visible *day and night*. A rear reflective marker
may be used in place of the official end
marker board on rear projections.

Attendants
Where an attendant is required his duty is to
ensure that the load does not inconvenience
or cause danger to other road users. It is not
necessary for him to hold a driving licence,
but he must have a thorough understanding
of the load and of the vehicle's capabilities.
(See also Chapter 16).

P: Police notification
Prior to the movement of any load marked
with the 'P' symbol, two days' notice must be
given to the chief of police in each area
through which the load is to pass. This is
normally dealt with by the Traffic Office,

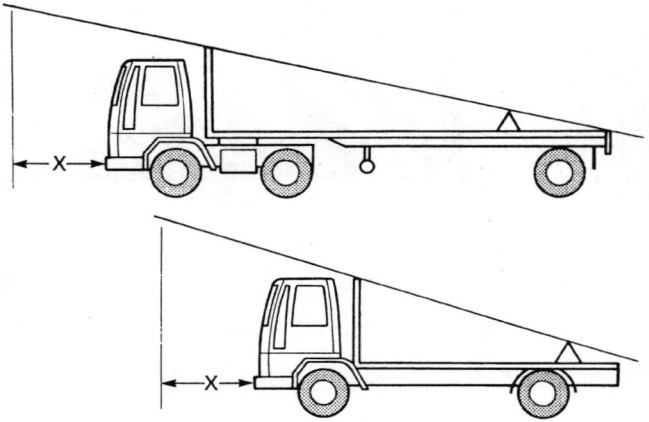

Up to 1.83 m (6'): No special requirement

1.83 m
to 3.05 m (10'):

Over 3.05 m:

 **P**

More than 4.5 m:

 Additional

X = forward projections
are measured from the
extreme front of the
vehicle to the tip of the
projecting load

and any objections (such as a low bridge, or
Figure 14

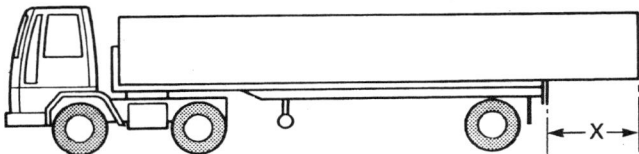

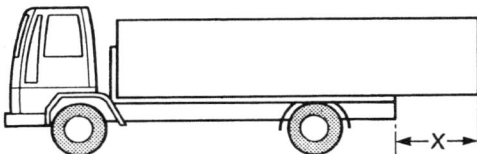

Up to 1.07 m (3'6"): No special requirement

1.07 m to 1.83 m (6'): Must be made clearly visible (CV)

1.83 m
to 3.05 m (10'):

Over 2.05 m:   P

More than 5.1 m:  Additional

X = rear projections are
measured from the
extreme rear of the
vehicle to the tip of the
projecting load. All
marker boards must
be adequately illum-
inated at night.

Forward Projections
Figure 15

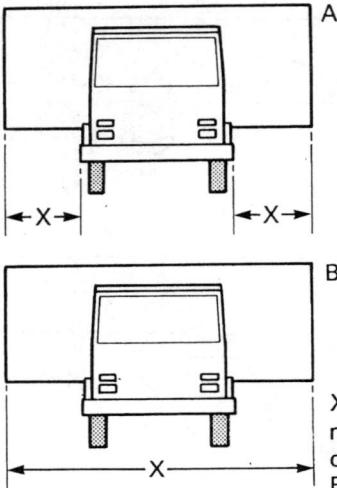

X = side projections are measured from: A. One end of the projection to the other; B. The side of the vehicle to the side of the projection

Rear Projections

Loads may not project more than 305 mm (1') on either side of a vehicle or trailer and the overall width of a vehicle and its load must not exceed 2.9 m (9' 6"). Where loads exceed 2.9 m, the following requirements must be observed:

Up to 2.9 m (9' 6"): Normal running; no special requirements

2.9 m
to 3.5 m (11' 6")   **P**

3.5 m to 4.3 m (14')   [person symbol] **P**

Note: Loads over 2.9 m are not permitted unless they are loose agricultural produce or are indivisible. Loads exceeding 4.3 m are not permitted uner normal Construction & Use Regulations, but are subject to Special Types regulations

Figure 16
Side Projections

a procession using the same route) will be explained to the company.

CV: Clearly visible
Where the law requires that the load be made clearly visible, and an official end marker board is not readily available, you may use highly coloured rags or paint.

9
Drinking and Driving

Preliminary Screening Test
Breath Test On An Alcolmeter
Positive Screening
Police Station Procedure
Penalties

Under the provisions of the Road Traffic
Act 1988 it is an offence to drive, or attempt
to drive, or to be in charge of a motor vehicle
when the proportion of alcohol in the blood
exceeds the prescribed limit of 80
milligrammes of alcohol per 100 millilitres of
blood.

The introduction of the alcolmeter range
of breath/blood alcohol analysers from 1981
means that a driver who has provided a
positive sample of breath on the main
alcolmeter 3000 (as illustrated in Figure 17)
can be convicted on the evidence of that
machine only (or of its equivalent).

The prescribed limits of alcohol in the
body as indicated by the breath testing
machine are 35 microgrammes of alcohol per
100 millilitres of blood.

The limits of blood to alcohol and urine to
alcohol are: 80 milligrammes of alcohol per
100 millilitres of blood.

Figure 17
The Alcolmeter for Screening Purposes Only

PRELIMINARY SCREENING TEST

A police officer may require a person to take a preliminary screening test by the side of the road if the officer concerned:
1. Has been called to the scene of an accident
2. Has reasonable grounds to suspect the driver of having committed a moving or stationary traffic offence, or
3. Has reasonable cause to suspect a driver of having alcohol in his body.

The equipment used by the police for the purpose of roadside screening can be either:

- the traditional breathalyser, or
- the alcolmeter breath testing machine.

BREATH TEST ON AN ALCOLMETER

Where the alcolmeter is used a person will be required to blow a sample of breath into the small hand-operated breath analyser. Should the test prove positive the alcolmeter will register accordingly, and that person will be arrested and the results can be used in court to secure a conviction.

POSITIVE SCREENING

A driver who provides a positive sample of breath, or a driver who refuses to give a sample of breath, or a driver who *fails* to provide a sample of breath will be arrested and taken to the nearest police station equipped with the main breath testing machine – the intoximeter 3000 or its equivalent.

POLICE STATION PROCEDURE

On arrival at the designated police station the driver will be escorted to the charge room where the alleged offence will be explained, and the driver's attention will be drawn to the 'Notice to Persons in Custody'. This provides that where a person has been arrested and is held in custody in a police station he shall be entitled to have intimation of his arrest and of the place where he is being held sent to one person reasonably named by him without delay or with no more delay than is necessary.

Breath analyser room

At this stage in the proceedings the driver (who now becomes the prisoner) is handed over to the official police intoximeter operator who will take the prisoner to the breath analyser room. There he will be warned once again of the offence and told what is involved in taking the breath test via the main intoximeter 3000.

When he understands the requirements, the prisoner will be asked to provide two specimens of breath.

Note: Failure or refusal to provide the required specimen without reasonable excuse will render the prisoner liable to imprisonment, a fine and disqualification.

The intoximeter test

For the purposes of the test the police operator will programme the machine to recalibrate and clear itself of any possible traces of alcohol before the test. The visual display unit will indicate that this process is complete.

The police operator will then feed into the machine his personal details (rank, police number, name and his personal operator's number). He will then feed the machine with the prisoner's personal details (full name, date of birth, date of test).

Only when the machine has accepted the information will it indicate on the visual display unit that the test may begin.

First test

The prisoner will be asked to blow into the machine through a sterile mouthpiece. He must continue to blow until told to stop. When the first test has been completed the intoximeter will clear and recalibrate itself before the next test. Only when the machine has cleared itself will it indicate that the second test may begin.

Second test
When the prisoner has blown into the
intoximeter for the second time the test is
complete. The machine will clear and
recalibrate itself again and continue to
analyse both breath samples. The results of
the test will be produced on the computer
print-out.

Print-out
The print-out (as illustrated in Figure 18) is
produced in triplicate, each copy giving the
results of both tests. The prisoner and the
police operator must sign all three copies of
the print-out in the presence of each other.
The triplicate copies are then divided and
one copy is given to the prisoner. Another
copy is put on case file, and the third copy is
kept in the station log records.

Legal status
Although the legal limit is 35 microgrammes
of alcohol per 100 millilitres of blood, the
subject who provides the two samples of
breath will be prosecuted on the *lower*
reading. If, for example, the first test showed
a reading of 50 microgrammes of alcohol per
100 millilitres of breath, and the second test
showed 48 microgrammes per 100 millilitres,
the *lower* reading of the two will be put
forward as evidence.

Furthermore, should the lower reading be
between 35 and 40 the police will *not*
normally take action, but should the reading
be between 41 and 50 the subject will be
given the option of having that analysis
replaced by a specimen of blood or urine.

Figure 18 (opposite)
The Intoximeter 3000 Print-Out
It is on the evidence of this print-out that a person could
be convicted in a court of law for any offence involving
drinking and driving.

```
TEST RECORD

LION INTOX.3000.5420 ————— Make of Machine
GMP ————————————————— Police force – Station

      DIVISION ————————— Police Division

WED JAN 08.1997 ——————— Day/Date

  SUBJECT NAME = ————————— Subject's full name

  DOB = 00 —————————————— Subject's date of birth


----------------- —————————— Subject's signature
SIGNATURE


TEST   UG%   TIME ————————— Test Microgrammes    Time

STD    33    09.48GMT ————— Reading              Time
BLK    0     09.49GMT ————— Zero                 Time

ONE    0     09.50GMT ————— Test No. 1           Time
BLK    0     09.50GMT ————— Reading              Time
TWO    0     09.52GMT ————— Test No. 2           Time
BLK    0     09.52GMT ————— Reading              Time

STD    34    09.53GMT ————— Reading              Time

                           Operator's name (Police)
  OPERATOR NAME = —————————

                           Endorsement of the text
I CERTIFY THAT IN THIS —————
STATEMENT, READING
ONE RELATES TO THE
FIRST SPECIMEN OF
BREATH PROVIDED BY THE
SUBJECT NAMED ABOVE,
AND READING TWO TO THE
SECOND, AT THE DATE
AND      TIMES     SHOWN
HEREIN.


                           Police operator's signature
----------------- —————————— Should the lower
SIGNATURE
```

reading be more than 50 the police *will* automatically charge the subject with the offence.

Detention
A person will *not* be detained at the police station provided there is no likelihood of their driving.

PENALTIES

The possible penalties for drinking and driving are severe. They could include
- death or injury to yourself or others
- disqualification from driving for *one year*
- car insurance trebled when you return to driving
- up to £5,000 fine, six months in prison, or *both*
- loss of job perhaps leading to indefinite unemployment and serious financial problems
- social stigma.

The drink/driving laws are complicated. Drinking and driving is simple – it can have devastating effects on both you and your family.

The courts at all levels have hardened their views, and now very much reflect the public attitude that drink-driving is socially unacceptable.

There are two aspects of drink-driving which should be emphasised. First, the length of time it takes for blood alcohol levels to drop below the legal limits for driving.

The second aspect is of concern to drivers of long distance commercial vehicles. If a driver is found to be drunk **in charge of** a stationary vehicle, this is also an offence. In such an instance disqualification is discretionary but in many cases the Court **will** disqualify.

There is only one sensible course –
DON'T DRINK AND DRIVE.

10
Motorway
Regulations

Motorways
Driving In Fog
Emergency Telephones
Speed Limits
Overhead Warning Signals On Motorways
Roadside Warning Signals On Motorways
Diversions From Motorways

MOTORWAYS

The first motorway section ever to be
completed in the UK was the M6 Section
between Fulwood and Preston. From this
beginning motorways have developed into
vast networks and motorway links. There are
a number of special rules that apply to
motorways and motorway driving. For
example the following vehicles may not use
the third lane of a three lane motorway:

- Large Goods Vehicles – artic and rigid
- Passenger carrying vehicles exceeding
  12 m
- Any vehicle drawing a trailer.

Large goods vehicles being driven in the
outside lane of a two lane motorway which
becomes the outer or extreme lane – right
hand lane of a three lane (or more)
motorway, ie at intersections, may continue
in that lane until such time as it becomes safe
to move into middle/adjacent lane on the
near side.

Road works on motorways

When about to encounter road works or
contraflow systems drivers of large goods
vehicles must keep to the left unless signs
indicate otherwise.

Obstruction on the carriageway

If anything falls from your vehicle or any
other vehicle that could be dangerous to
other motorway users, stop at the next
emergency telephone and inform the motorway police who will respond immediately, do
not attempt to recover it yourself.

Picking up passengers

It is an offence to pick up or set down
hitchhikers on any part of a motorway
including the slip roads.

DRIVING IN FOG

Before driving in fog, consider if your
journey is essential. If it is, allow extra time.
Make sure your windscreen, windows and
lights are clean and that all your lights
(including brake lights) are working.
When driving in fog:

- See and be seen. If you cannot see clearly
  use dipped headlights. Use front or rear
  fog lights if visibility is seriously reduced,
  but switch them off when visibility
  improves. Use your windscreen wipers
  and demisters.
- Check your mirrors and slow down.
  Keep a safe distance behind the vehicle in
  front. You should always be able to pull
  up within the distance you can see clearly.
- Do not hand on to the tail lights of the
  vehicle in front; it gives a false sense of
  security. In thick fog, if you can see the
  vehicle in front you are probably too
  close, unless you are travelling very
  slowly.

- Be aware of your speed; you may be going much faster than you think. Do not accelerate to get away from a vehicle which is too close behind you. When you slow down, use your brakes so that your brake lights warn drivers behind you.
- When the word 'Fog' is shown on a roadside signal but the road appears to be clear, be prepared for a bank of fog or drifting smoke ahead. Fog can drift rapidly and is often patchy. Even if it seems to be clearing, you can suddenly find yourself back in thick fog.

EMERGENCY TELEPHONES

If the vehicle develops a problem, leave the motorway at the next exit or pull into a service area. If this cannot be done:
- Move onto the hard shoulder and stop
- Try to stop near an emergency telephone; these are located at one mile intervals
- Stop as far to the left as possible, switch on the hazard warning lights and side lights
- Leave the vehicle by the near side door and ensure any passengers do the same
- If the driver has to leave the vehicle in order to summon help, ensure that any passengers wait near the vehicle, but well away from the carriageway and hard shoulder
- Walk to the emergency telephone following the direction arrows on the marker posts (do not cross the carriageway)
- Emergency telephones are directly connected to the motorway police and are free to use
- Give as much detail to the police operator as possible ie name of driver, name of haulier, type of vehicle, location

on the motorway, type of assistance
required, ie tyre service, recovery service,
etc
- Return to the vehicle and wait for
assistance.

SPEED LIMITS

Speed limits are governed either by the road
or the vehicle; whichever is the lower applies.
Vehicles being driven on restricted roads are
generally restricted to 30 mph, unless there
are approved road signs which indicate
otherwise.

Speed limits can be monitored in a
number of ways including:

Radar
Radar is a hand-held device, which is able to
record the speed of an oncoming vehicle by
calculating the distance covered as identified
by a diminishing beam which would be
pointed at a vehicle.

Vascar
Vascar is a device which is installed in police
vehicles and is able to measure the distance
between two points. It can calculate the time
travelled by a vehicle and its speed.

Gatso camera
Cameras may be installed at the roadside or
at strategic places on the motorways. These
are able to photograph vehicles exceeding the
speed limits or vehicles passing traffic lights
on red. A camera device can trigger a
mechanism to produce photographic
evidence showing the date and time of an
offence, in addition to the position of a
vehicle and the speed at which it was
travelling.

Speed limits	Built-up Areas	Elsewhere		Motorways
		Single carriage-ways	Dual carriage-ways	
Type of Vehicle	**MPH**	**MPH**	**MPH**	**MPH**
Cars (including car-derived vans and motorcycles)	**30**	**60**	**70**	**70**
Cars towing Caravans or Trailers (including car-derived vans and motorcycles)	**30**	**50**	**60**	**60**
Buses and Coaches (not exceeding 12 metres in overall length)	**30**	**50**	**60**	**70**
Goods Vehicles (not exceeding 7.5 tonnes maximum laden weight)	**30**	**50**	**60**	**70** (60 if articulated or towing a trailer)
Heavy Goods Vehicles (Exceeding 7.5 tonnes maximum laden weight)	**30**	**40**	**50**	**60**

Figure 19

OVERHEAD WARNING SIGNALS ON MOTORWAYS

Flashing red signal.
Do not proceed any further in this lane

Change lane

Leave the motorway at the next exit

Temporary maximum speed

End of restriction

Figure 20

ROADSIDE WARNING SIGNALS ON MOTORWAYS

These signals are situated on the central reserve, at intervals of not more than two miles. They apply to all lanes.

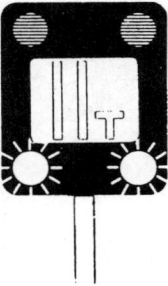

Lane closed ahead   Temporary maximum speed   End of restriction

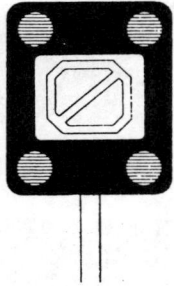

Figure 21

DIVERSIONS FROM MOTORWAYS

It is sometimes necessary, because of emergencies, to close either one or both carriageways of a motorway. When this is done, special signs will sometimes be shown to tell drivers on the motorway (or those wanting to join it) about the alternative route they should follow. This route will guide them back to the motorway beyond the point of closure.

The diversion sign could be part of an existing direction sign or a separate sign on its own. If the diversion route is easy to identify and follow, the signs will show only its route number (for example A38 via Bridgewater) and there will be no additional diversion signs along this route. This is a typical sign.

Figure 22

Sometimes there are too many route numbers to show on the diversion sign, or the alternative route may not have a route number. In such cases the diversion route will be indicated by symbols of various shapes shown on the diversion sign and along the recommended route. The symbol may be a rectangle, a circle, a triangle or a diamond and each of these may be either black, or yellow with a thick black border, for example:

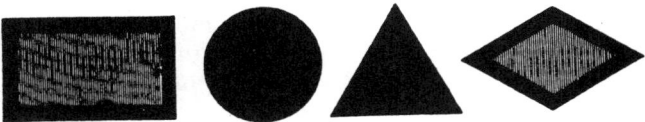

On a motorway diversion sign the symbol will be like this:

Figure 23

Along the diversion (alternative) route,
drivers should look for the symbol which will
be shown on the traffic direction signs which
will guide them back to the motorway.
Along the route, the symbols will be like this:

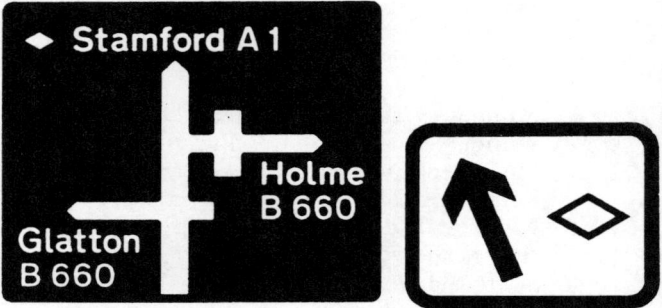

Figure 24

Where two or more diversion routes cross or
follow the same length of road, different
symbols will be used to avoid confusion.
Therefore, if you should be diverted from a
motorway by a sign which shows one of
these symbols, keep a good look out for the
coloured symbol along the alternative route
and always follow the correct symbol.

11
Accident
Procedure

A Checklist

All truck drivers hope that they will never be
involved in a road accident. It is only when
such an accident occurs that it becomes
apparent that accident procedures may have
been neglected as regards information to be
given and information to be obtained.

It is very important that *all* necessary
information is gathered *at the time of the
accident* so that the true facts can be
analysed and determined at a later stage.

The legal requirements
In the event of an accident which causes
injury to another person or an animal, or
damage to another vehicle or other property,
as defined in the Highway Code, the law
requires the driver to stop his vehicle and
give his own name and address and the
registration number of the vehicle to anyone
having reasonable grounds for requiring this
information. If it is not possible to give this
information at the time of the accident, the
accident must be reported to the police as
soon as is reasonably practicable, and in any
event, within 24 hours.

In the case of injury to another person, the
law requires that the insurance certificate be
shown to anyone having reasonable grounds
for requiring it – including the police – at the
time of the accident. If this is not possible,
the accident must be reported to the police as

soon as possible or within 24 hours. If it is not possible to produce the insurance certificate at this stage, it must be produced within seven days thereafter at any police station.

Essential information

In addition to the legal requirements needed at the scene of an accident, other pieces of essential information may be required by insurance companies, vehicle owners, or, indeed, the police. This will be information such as:

- the name and address of any reliable witness
- an accurate description of any damage
- the exact location of the accident
- a sketch outlining the position(s) of vehicle(s) involved, showing approximate distances, widths of roads and their priority, pedestrian crossings, traffic lights, road signs, speed limits, etc
- the condition of the road surface, eg wet or icy
- the visibility at the time of the accident, eg fog or bright sunshine
- the cause of the accident (but do not admit any responsibility – this is usually decided by the authorities at a later stage)
- the weight of the load, plated gross weight
- the date and time of the accident
- whether any warning was given (if the police were involved)
- the number of the police officer concerned
- the police station concerned
- the speed of the vehicles.

Accident forms

Accident forms, usually supplied by the insurance company or the vehicle owner, must inevitably be completed. It is always advisable to carry such a form so that, in the event of an accident, it can be completed at that time. Thus, a ready-made checklist of essential information is available, ensuring that nothing is forgotten. The form need only be completed roughly at the time of the accident and copied out more clearly at a later stage.

A CHECKLIST

- Eliminate danger to casualties
- Organise emergency telephone call
- Administer first aid if necessary
- Warn other traffic
- Take the name and address of any reliable witness
- Make a note of the date and time of accident
- Note damage to vehicles, exact location of accident and driving/road conditions
- Note weight of load, plated gross weight, speed of vehicle involved
- Do everything possible to co-operate with the emergency services but do *not* admit any responsibility – this will be decided by the authorities at a later stage.

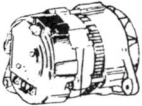
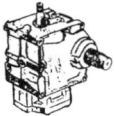

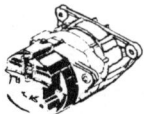

12
The Police and the Department of Transport

Police Powers And Your Rights
The Role Of The Department Of Transport

POLICE POWERS AND YOUR RIGHTS

A police officer in uniform is entitled to stop any vehicle on the road. You do not need to have committed an offence. If you fail to stop, or if you stop and then drive away before instructed to by the police, you commit an offence. It is also an offence to fail to conform to traffic directions given by police officers.

Police powers

The police are empowered to question any person suspected of having committed an offence, or suspected of having information useful to a policy enquiry. All police officers may report drivers for prosecution, or may arrest in serious cases. Police officers are also entitled to require the production of certain documents, to examine vehicles, and to require them to be weighed.

Documents

You must produce, if requested by a police officer, the following documents:

- Driving licence
- Evidence of insurance
- MOT test certificate or goods vehicle test certificate.

If you do not have them with you, you must produce them *at a police station of your choice* within seven days.

The police may also wish to examine documents carried on the vehicle:

- vehicle excise license
- vehicle identity disc
- manufacturer's plate
- ministry plate
- trailer test date disc
- trailer plate.

Excessive loads

A vehicle is overweight if it exceeds the weights specified on either the manufacturer's plate *or* on the ministry plate. Weights on ministry plates never exceed those on the manufacturer's plate.

If a police officer is authorised, he may on production of his authority, require a driver to allow the vehicle or trailer to be weighed.

- You commit an offence if you refuse
- An authorised police officer may issue a notice preventing an overweight vehicle from being drive on the road
- Not all police officers are authorised to have vehicles weighed
- If your vehicle is required to travel more than five miles to a weighbridge, and is found to be *within* the legal weights, you may claim costs from the Highway Authority (*not* the police).

Dangerous vehicles

Vehicles can be dangerous in three ways:

- Poor maintenance of vehicle and accessories
- Insecure load
- Unsuitability for use to which it is being put.

The vehicle must be safe for every journey. It is no defence to produce a current test certificate for the vehicle. In many cases both owner *and* driver are responsible.

THE ROLE OF THE DEPARTMENT OF TRANSPORT

The Department of Transport plays an important role in the administration of road transport but in general terms the Department's involvement with the driver revolves around his hours of work, keeping of records and the condition of his vehicle. To ensure that goods-carrying vehicles are properly maintained in a fit and serviceable condition the Department's Vehicle Inspectorate (VI) carries out various checks and inspections: *annual tests, roadside checks* and *vehicle checks on company premises.*

Annual test
Most large goods vehicles, articulated vehicles, rigid goods vehicles and trailers are required to be tested annually at Department of Transport goods vehicle testing stations. The test of the vehicle and its major components is very thorough and is completed in stages. A full test will normally take about 45 minutes during which time the driver is required to assist the examiner by operating certain controls and moving the vehicle when required. In total about 60 items will be looked at, checked or tested, a list of which appears opposite.

Roadside check
Roadside checks are usually conducted on wide, quiet roads just off major trunk roads, or in large lay-bys with enough room to accommodate several large vehicles at any one time.

The Vehicle Inspectorate examiners

Annual test checklist

- Position of legal plate
- Details of legal plate
- Smoke emission*
- Road wheels and hubs
- Size and type of tyres and their condition
- Bumper bars
- Spare wheel carrier
- Trailer couplings*
- Trailer landing legs
- Condition of wings
- Cab doors, floor and steps*
- Driving seat*
- Security and condition of body
- Mirrors*
- View to front*
- Condition of glass*
- Windscreen wipers and washers*
- Tachometer
- Audible warning*
- Driving controls*
- Play at steering wheel*
- Steering wheel*
- Steering column*
- Air/vacuum warning*
- Build-up of air/vacuum*
- Hand levers controlling mechanical braking systems*
- Service brake pedal*
- Service brake operation*
- Hand-operated air/vacuum control valves*
- Condition of chassis
- Electrical wiring and equipment
- Engine mountings*
- Fuel tank system*
- Exhaust system*
- Condition of spring pins and bushes
- Condition of suspension units
- Attachment of suspension units
- Shock absorbers
- Oil leaks*
- Stub axles/wheel bearings
- Steering linkage
- Steering box*
- Power steering*
- Transmission*
- Mechanical brake components
- Brake wheel units
- Brake pipes/master cylinder/reservoirs/valves/connections
- Rear markings
- Obligatory front lamps*
- Obligatory rear lamps
- Obligatory reflectors, side reflectors (on long vehicles and trailers)
- Functioning of direction indicators
- Vertical aim of headlights*
- Obligatory headlamps (position and function)
- Obligatory stoplamps (position and function)
- Trailer parking brake
- Maintenance of service brake
- Maintenance of secondary brake
- Maintenance of parking brake

* This does not apply to trailers

are not permitted or authorised to stop traffic. A police officer in uniform will therefore be in attendance at all roadside checks for the sole purpose of directing traffic chosen for examination into the appropriate position where the examination will take place.

Vehicle examiners have only a limited amount of equipment on these checks so their examination of vehicles is somewhat restricted. Nevertheless, the examiners are usually very experienced and extremely thorough. Minor defects will be pointed out and the driver will be given the opportunity to put them right. However, should a defect prove to be more serious the appropriate PG notice will be issued. A record of the inspection is provided as evidence of the roadside check.

Another type of roadside check is the one which is conducted by the Department of Transport vehicle examiners but without the assistance of the police to stop vehicles for the examination. On production of the examiner's authority card any vehicle examiner may approach a stationary vehicle, for example in a car park, and conduct a vehicle examination or request the driver to move the vehicle to a suitable place not more than one mile away where the vehicle can be examined.

Vehicles checks on site
Not all vehicle examinations conducted by the Vehicle Inspectorate are carried out at the side of the road or at official test stations. Authorised inspectors may enter premises where vehicles are kept and conduct vehicle examinations there and then on site. In order to do this, however, the owner's consent must be obtained beforehand or he must be given at least 48 hours' notice of such an examination.

Drivers' hours and records

Vehicle examiners are not normally authorised to carry out inspection of the driver's record chart (the tachograph) although a police officer may ask to see it. However, the Vehicle Inspectorate may assign a traffic examiner solely for that purpose who will accompany other officials, ie police and/or vehicle examiners, when they are carrying out roadside checks or checks on vehicles at premises.

Silent check

Silent checks are conducted by the Vehicle Inspectorate when it suspects that certain vehicles are being driven over the hours limit. The suspect vehicles are clocked to be in a certain place at a certain time and are checked again and again at other locations. Where there appears to be time/mile ratio discrepancy examining officers will usually make a point of calling on the suspected offender to carry out a full or part inspection of the driver's hours and records. Full details of drivers' hours can be found in Chapter 7.

13
Carrying Passengers and Giving Lifts

Passenger Liability
Company Law

Although the law of the land does not extend to the picking up of passengers by drivers of commercial vehicles, company law and insurance certainly do.

It is the responsibility of all employers to ensure that their vehicles are properly insured and the driver should know the legal position with regard to insurance. No motor vehicle may be used on a public road without an insurance policy which covers at least third party and passenger liability.

PASSENGER LIABILITY

In the event of an accident involving personal injury to a third party the certificate of insurance must be produced. Any instruction by the employer *not* to carry passengers *must be strictly observed*. Failure to do so may affect the insurance position in the event of an accident in which a passenger was injured.

COMPANY LAW

Because of the increasing number of problems which are associated with drivers picking up hitchhikers, more and more transport companies are forbidding the carriage of passengers in their vehicles. This is known as company policy and failure to adhere to it may and often does result in immediate dismissal of the offending driver.

Warning or disclaimer plates
To ensure that drivers are aware that carrying passengers in their vehicles is a dismissible offence, warning plates and/or disclaimer plates can often be found in drivers' rest rooms or fixed into the cab of the vehicle, usually in a prominent position where they can be clearly seen by both the driver and the potential passenger
Examples of disclaimers are:

**'The carrying of passengers in this vehicle is strictly forbidden.
Drivers failing to comply with this instruction are liable to instant dismissal.'**

'Passengers who travel in this vehicle do so entirely at their own risk.'

'No passengers must be carried in this vehicle.'

Several problems can occur for the truck driver giving a hitchhiker a lift. For example, a driver carrying an expensive load runs the risk of hi-jack. All drivers also run the risk of being involved in a vehicle accident where the passenger may be injured. One possible accident can occur when a passenger, who is not used to climbing in and out of heavy goods vehicles, slips and falls out of the vehicle. A good deed can sometimes turn into a nightmare if a hitch-hiker passenger receives an injury, no matter how slight, and later reports it to the appropriate authority in order to reap compensation.

Drivers who choose to ignore these warnings do so at their own peril and are foolish to think 'it won't happen to me'.

14
Vehicle and Load Security

INTRODUCTION TO SAFE LOADING

If you consider the different types of cargoes that are carried on our roads the list would be endless, and it must be stressed that it is the driver who is responsible for that cargo, irrespective of who loaded the vehicle.

Loading vehicles is a skilled job and drivers must ask themselves four questions before loading begins:

1. Have you got the right vehicle for the load?
2. Are you sure that the maximum permitted weight of any one axle will not be exceeded?
3. Are you sure that the maximum gross plated weight of the vehicle will not be exceeded?
4. Are you sure that the load can be evenly distributed, packed and contained securely on the vehicle platform?

No matter how experienced a driver may be at loading different cargoes, there will always be an occasion when he/she is confronted with a cargo that he/she has not loaded before – in that case, ask.

The suceeding test offers only a few suggestions on how to load certain cargoes safely onto the vehicle platform.

BLOCKING AND BINDING INTERLOCKING CARGOES

When loading cartons or bags, Blocking and Binding is an accepted method that will not only offer interlocking stability of the cargo, it will also enable a quick and reliable count at any stage of the loading operation.

Place the first leg in a gunshot position as indicated in Figure 25 and complete the leg by placing the remaining cargo in a rolling position as indicated in Figure 26, this will give a leg of 8 cartons or bags.

Note: Subject to the size of the carton or bag the carton would be more or less.

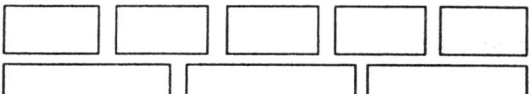

Figure 25

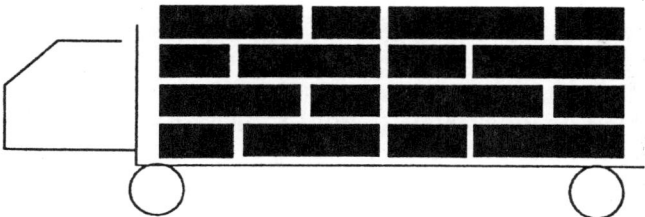

Figure 26

Repeating the process in reverse will build a safe and secure load.

NON-INTERLOCKING CARGOES

When loading square cartons in any number
you do not have the natural facility to
interlock them with each other.

In this case a simple sheet of paper placed
between each row of cartons as illustrated in
Figure 27 will help stabilise the load and also
help to reduce the risk of the cartons fanning
at the top.

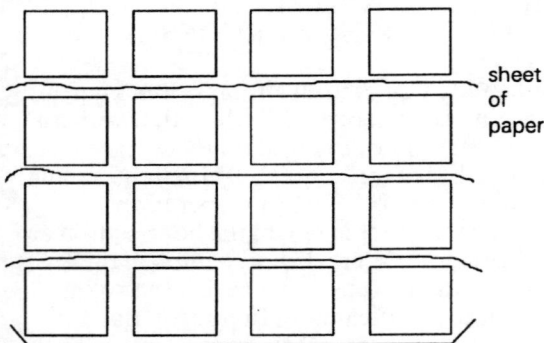

sheet
of
paper

Figure 27

USING SKIDS

When loading steel or concrete, for example,
it is important not to place the cargo directly
on the vehicle platform unless special
provision has been made for the lifting and
lowering.

Place the skids across the platform of the
vehicle, as indicated in Figure 28, place the
cargo on the skids and repeat the process,
ensuring that all skids are in line, this will
ensure all weight stress points are supported.

Note: When securing the cargo using straps
or chains make sure the securing devices are
placed in line with the skids as indicated X.

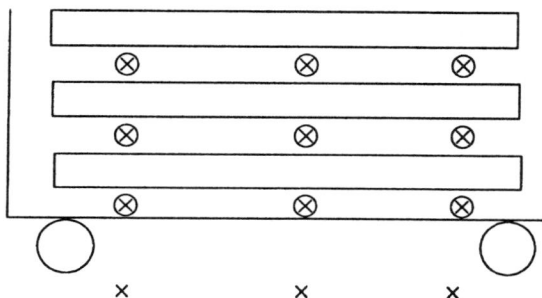

Figure 28

OIL DRUMS OR BARRELS

When loading oil drums or barrels, ensure
that each drum makes contact with the other
as indicated in Figure 29. This will offer
maximum coverage of the vehicle platform
and reduce the risk of the cargo sliding
about. In securing the drums a fly sheet is
normally sufficient, placed across the top of
the drums or barrels and secured with ropes
as in Figure 30.

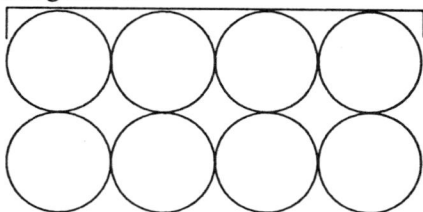

Figure 29

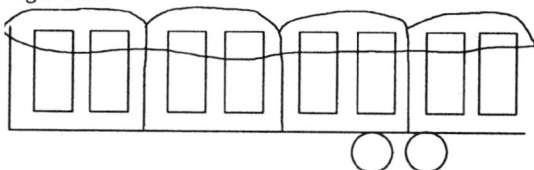

Rope between every other barrel and cross at rear
Figure 30

ROPING AND SHEETING

Before you start there are four questions to ask again:
- Have you got the right vehicle for the right load?
- Are you *sure* that the maximum gross plated weight of the vehicle is not exceeded?
- Are you *sure* that the maximum permitted weight of any one axle is not exceeded?
- Are satisfied that the load is evenly distributed, packed and contained securely on the vehicle?

You can be fined up to *£5000* for infringements of loading and weight regulations.

Remember: this may affect your LGV licence.

Sheets
Lorry sheets protect your load. Tears and other faults should be reported as soon as possible because, if they are not repaired, the safety of your load may be threatened.

Folding the sheet
When folding sheets the top side must face upwards. Make sure, then, that the area where you are working is clean as the side that will be in contact with the load will be facing the floor. Then:
- Open sheet fully
- Fold cover A and A1 to form a neat crease about 60 cm (2 ft) from the centre line
- Bring the ends of the sheet A and A1 back in line with the edge of the newly formed crease
- Repeat on other side of sheet
- Fold the two ends to the middle
- Extract the trapped air by walking over the sheet

- Roll tightly.

See Figure 31 for an illustration of folding the sheet.

Note: If you fold your sheets correctly they will fall evenly over the load the next time you use them.

Sheeting the load
The way you sheet a load is, of course, largely dictated by its shape. Whatever the shape *always take care* – never walk backwards towards the edge of the load!

For *square* loads:
- Place folded sheet on top of load and unroll it
- Allow sheet ends to fall almost in line with bottom of chock rail
- Gather up the sheet side and throw it out over one side of the load
- Repeat on other side.

Figure 32 illustrates this.

For *longer* loads:
- Use two sheets to cover load properly
- *always* put back sheet on first
- Tie off.

See Figure 33 for an example. Figure 34 demonstrates enveloping the end of the load.

Using the fly sheet
A fly sheet is generally longer and narrower than normal lorry sheets. When your load is *completely roped and sheeted* spread the fly sheet along the top allowing it to drape over all sides, and then secure it with its own ties. The fly sheet prevents pools of water gathering on top of the load and, no matter how tightly it is secured, when the vehicle is moving the air flow will prevent water settling (see Figure 35).

Using three sheets
If you have an exceptionally large load you

Centre line
of sheet

A

A1

Crease 2' from
centre line

Repeat the
procedure

Fold ends
into centre

Roll neatly

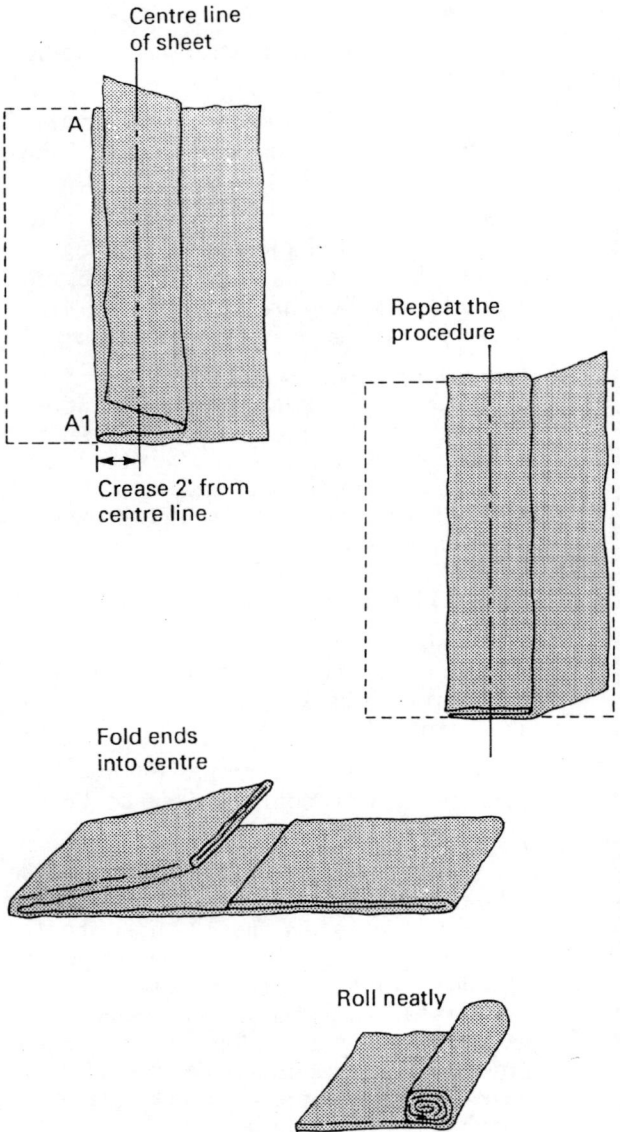

Figure 31
Folding the Sheet

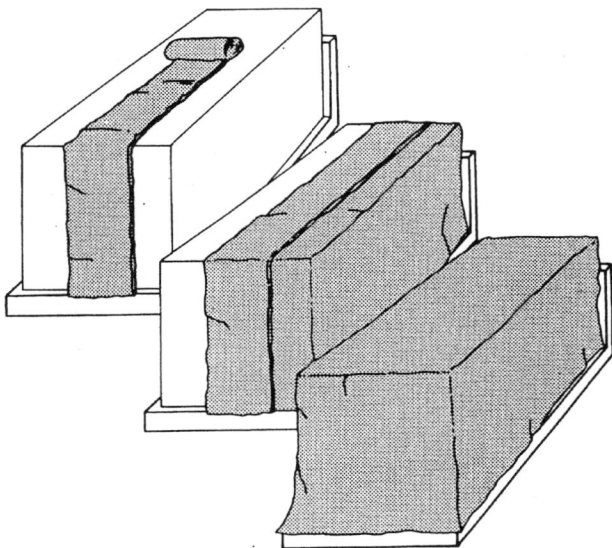

Figure 32
Using One Sheet

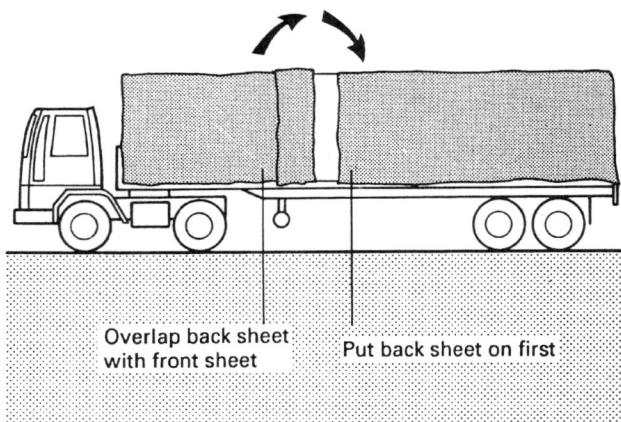

Overlap back sheet
with front sheet

Put back sheet on first

Figure 33
Using Two Sheets

When the sheet has been spread loosely over the load ensure that the sheet is in line with the bottom of the chock rail all round. Secure the front end of the sheet, then pull one corner of it round the back of the vehicle platform and tie off

Pull the other corner round and tie that end off, forming an envelope

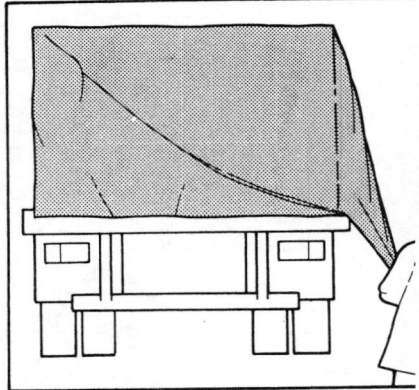

To complete the enveloping, pull out the middle of the sheet and square off the corners. In this way it will look neat, it will eliminate any water traps and allow the rain to run freely off the load

Figure 34
Enveloping

may need to use two large sheets and a fly
sheet to protect it.

- Put fly sheet along top of load and
  carefully drape it down the whole of one
  side
- Secure sheet firmly to top of load
- Drape *back* sheet over load and pull
  down on the side *opposite* the fly sheet
- Repeat with front sheet
- Tie sheet ends down all round
- Rope off (see Figure 37).

Sheet securing ties
- When you have draped the sheets loosely
  over the load use the ties which are fixed
  to the eyelet flaps to secure them
- You should complete the entire sheeting
  process *before* the load is roped
- Don't overtighten sheet ties with 'dolly'
  knots – you could tear the eyelet out of
  the flap
- Fold sheet under chock rail to prevent
  sheet from riding up, to minimise rope
  fray, and to increase the security of the
  load (see Figure 36).

Ropes
Ropes are usually 30 m, 45 m or 60 m (110,
150 or 200 ft) long depending on vehicle size.
The rope ends should be sealed with a back
splice and an eye splice (see Figure 38). Rope
which has been cut from a main coil will be
twisted and should be stretched before
folding.

Folding the rope
- Always begin with the stub end
- Folds should be even to prevent tangles
- Leave about 1–2 m (4ft) of rope at the
  end to secure coil (see Figure 39).

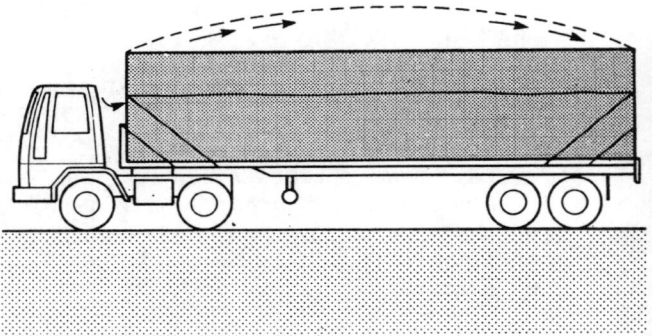

Figure 35
Using a Fly Sheet

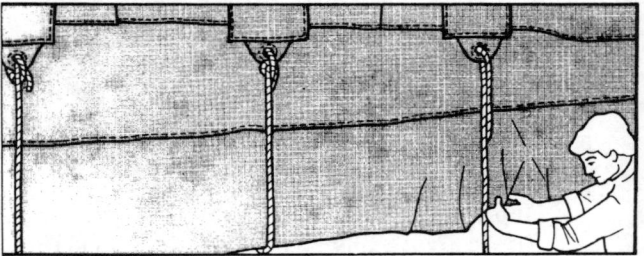

Figure 36
**Folding the
Sheet Under
the Chock Rail**

Fly sheet over one
side and tied down

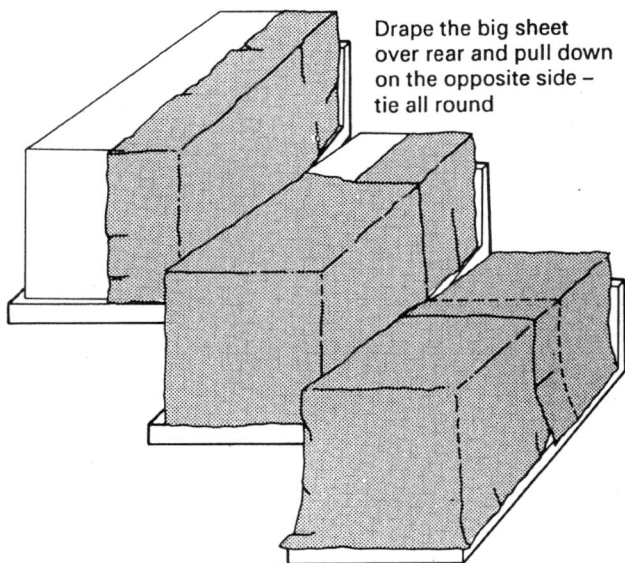

Drape the big sheet
over rear and pull down
on the opposite side –
tie all round

Drape the second big
sheet over the front,
overlapping the rear sheet -
tie all round before roping

Figure 37
Using Three Sheets

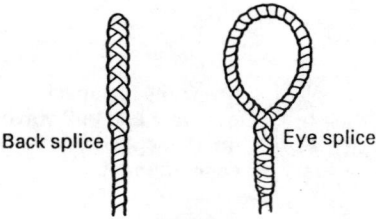

Back splice   Eye splice

Figure 38

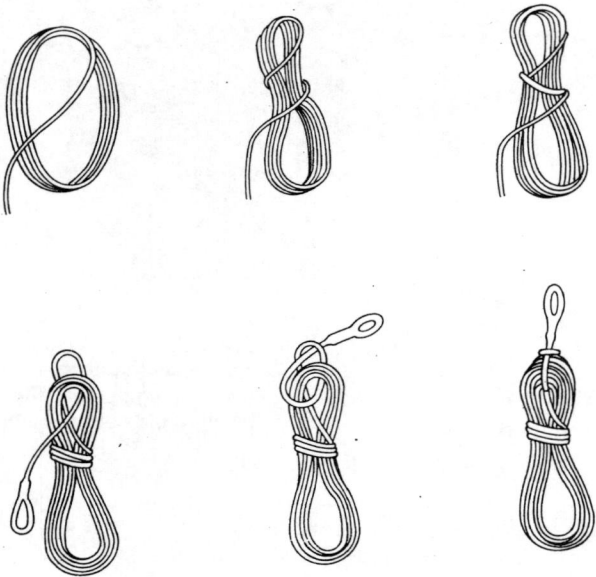

Figure 39
**Folding the
Rope**

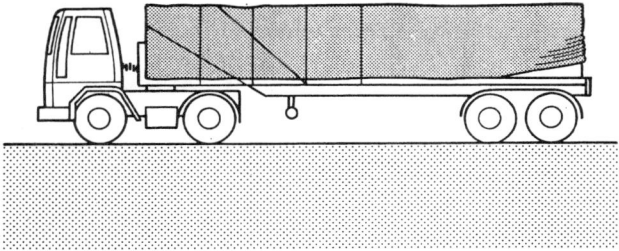

Figure 40

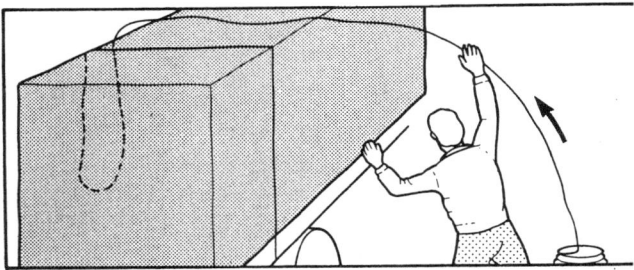

Figure 41
**Throwing the Rope
Over the Load**

Roping the load
As with sheeting, the way you rope the load
will be dictated by the type of load and by
the way the sheets have fallen.
- Take care that you don't so overtighten
  the ropes at one end of the load that the
  sheets at the other end ride up (see
  Figure 40)
- Hang the loop end of the rope on the
  rope hook and throw a small coil
  (enough to go over and back) over the
  load. *Make sure that it is clear on the
  other side!* (Figure 41)
- Tighten the rope on the other side with a
  'dolly' (see Figure 42), and lock it as
  shown in Figure 43

- Ropes should be generally cross the load between parallel rope hooks. Where this is impossible because a hook is badly damaged, or missing, use a spreader as shown in Figures 44 and 45
- You will probably need more than one rope. The two easiest methods of joining ropes are the *standard rope join* (Figure 46) and the *semi-dolly rope join* (Figure 47).

Chains and sylvesters
Chains and sylvesters are used to secure heavy equipment

Chains
Unlike ropes, chains should *not* be attached to rope hooks but should be placed around the chassis of the vehicle. *Take care not to interfere with vehicle components* – prop shafts, air tanks, trailer brakes, cables, etc.

Sylvesters
- Open sylvester, position on *offside* of load where possible, and attach to chain
- Check that chain is square around load, that it is not twisted, and that it is clear of vehicle components
- Pull sylvester arm down, ensuring that chain is firmly on its gripping claw (see Figure 48)
- *Take care* – the tension forces the sylvester to snap shut very quickly
- Wrap surplus chain around sylvester as an added safety measure.

Tubular steel bars
These are not normally recommended but some drivers find them an added protection, and they do provide additional leverage when you are closing or opening the sylvester arm. The two things to keep in mind at all times are:
- your own safety when using sylvesters
- the security of your load.

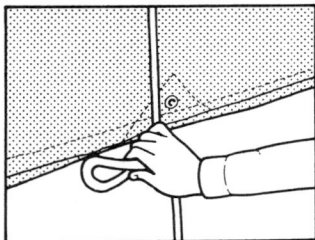

1. Tying the rope, or 'dolly', as it is better known, begins with the forming of a loop in line with the bottom of the chock rail

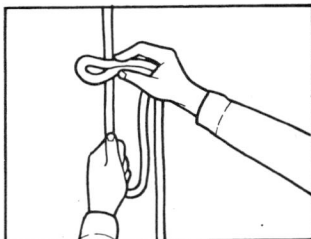

2. Place the loop about 2' up the rope

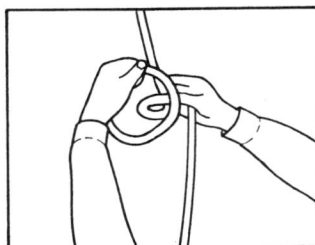

3. Wrap the rope around the loop

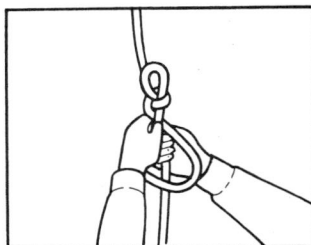

4. This will create a larger loop beneath the original. Hold firmly and twist the large loop anti-clockwise

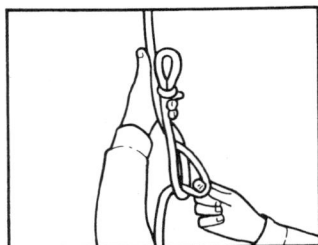

5. Still holding firmly with the left hand, pull the back rope through the newly formed loop

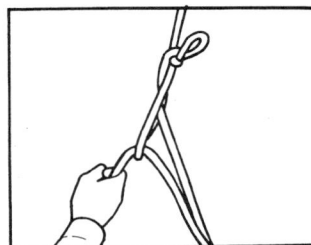

6. The dolly is now complete but should be held firmly. Finally, place the rope around the rope hook and pull tight

Figure 42
Tying the Rope

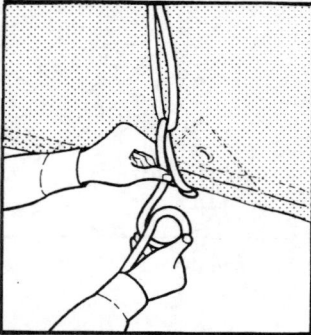

1. When the dolly knot is really tight it will need to be locked. Using the thumb of the left hand, press the rope into the space created by the hook and with the right hand form a loop

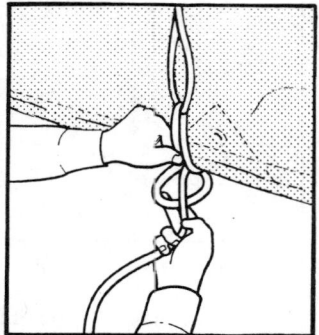

2. Place the loop around the rope hook, keeping the thumb firmly in position, and with the right hand pull tight to complete the lock

Figure 43
Locking the Dolly

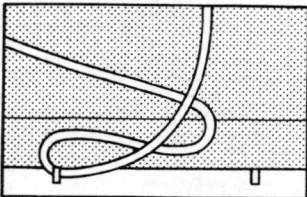

Place the rope round the hook and pass over the hook and round the back of the rope

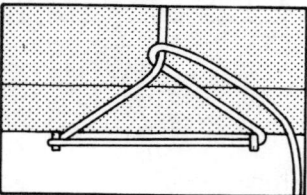

Take the rope on to the next hook and pull tight

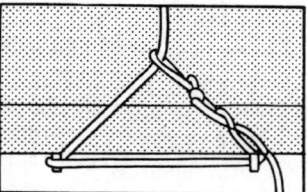

Make any final adjustments which might be required, then tie off using a normal dolly knot and locking hitch

Figure 44

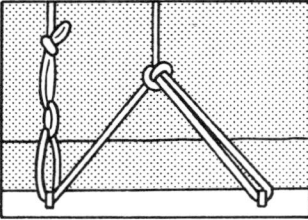

Where the rope comes from the adjacent rope hook it will be necessary to use a completely different type of spreader

Figure 45

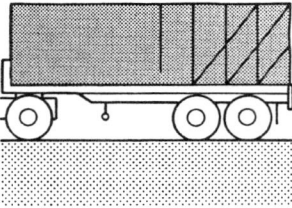

Where there is not enough rope to go over the load again but too much to waste, a standard rope joint will be needed

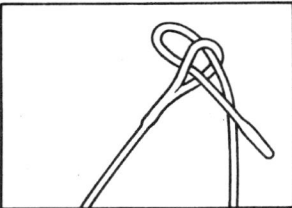

Using the loop end of the second rope and the stub end of the first, fold the stub of the first rope and place it through the loop end of the second

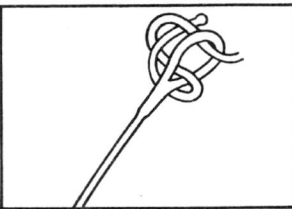

Pass the stub end of the first rope through the newly formed loop

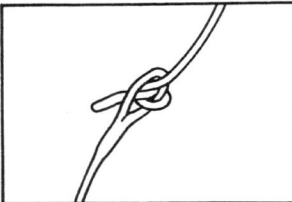

To completely join, pull tight. Not only will the join be strong enough to stand up to normal stress and strain when the rope is taut; it will also be easy to loosen when the strain is taken from the knot

Figure 46
Standard Rope Join

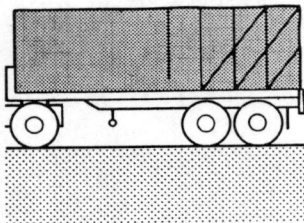

Where the rope falls slightly short of the hook a semi-dolly rope join will be needed

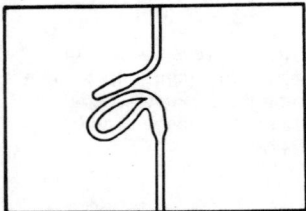

Using the stub end of the first rope to form part of the dolly and the loop end of the second rope to complete it, place the stub end of the first rope next to the loop end of the second

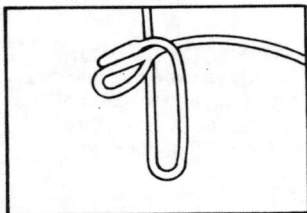

Holding the two ends firmly together, form a loop by placing the ends about 60 cm (2') up the rope and continue by tying a normal dolly knot as described in Figure 41

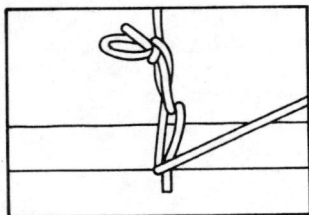

This will complete the join via the dolly which should then be pulled tight and locked

Figure 47
Semi-dolly Rope Join

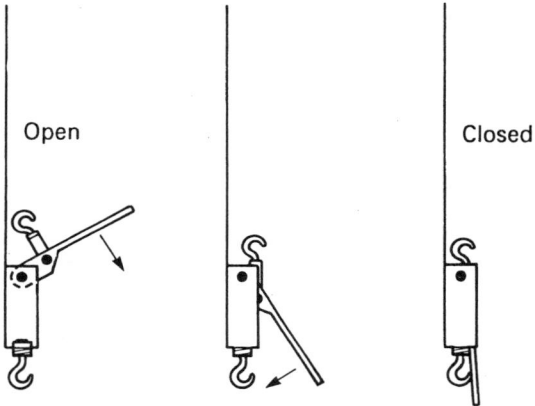

Figure 48
Sylvesters

CONTAINER SECURITY

Containers should only be carried on vehicles or trailers equipped with the appropriate twist locks designed to lock into the container body.

Whatever type of container is carried, all locking levers must be in the secured position during transit. Containers should not be carried on flatbed platform vehicles where there are no means of locking the container in position.

Never rely on the weight of the container and its contents to hold it in place on a flat deck. Ropes are totally inadequate to hold a typical sea-going container in place.

Skeletal vehicles or trailers have a main chassis-frame with outrigger supports into which the container can be locked.

TAUT LINERS

Taut Liners is another word for curtain-sided vehicles.

The manufacturers of vehicles fitted with curtain-side bodies are satisfied that a high degree of protection is given by the material used in their construction.

This does not relieve the driver of the responsibility for ensuring that the load is properly stowed and secured so that it will not move whilst in transit.

Take notice of warnings of adverse weather conditions broadcast on the radio – especially if your vehicle is empty.

Under such conditions it is often safer to secure both curtain sides at one end of the vehicle – thus cutting down the wind resistance and removing the likelihood of being blown over or off the road.

SAFETY AND SECURITY – A CHECKLIST

The following suggestions are not, of course, comprehensive, but none should ever be neglected.

Personal safety
- If you drive a vehicle of more than 3.5 tonnes which is fitted with seatbelts you *must* by law to wear the seatbelt. Seatbelts reduce the likelihood of serious injury. Bear in mind, too, that the courts favour drivers who wear seatbelts
- Wear close-fitting overalls as loose clothing can cause accidents. Take care, similarly, with rings, ID bracelets, etc
- Don't wear your hair in such a way that it interferes with vision
- Make sure that your footwear is appropriate to the type of work you are doing.

Safe loads

- You are responsible for the safety of your load – remember that your licence is at stake
- Be sure that your load is evenly distributed, within the total axle weight and within the total gross vehicle weight – this too is *your responsibility*.

Fire safety

- Take care with oil spillage and waste oil in the workshop
- *Do not smoke* – except in designated smoking areas
- Don't park in front of fire exits or escapes
- If you have a tyre fire call the fire service – only a continuous jet of water will put the fire out and keep it out
- If you have a cab fire
  - evacuate the cab
  - try to fight the fire upwind and avoid inhaling smoke. Hot smoke in the lungs is often fatal
  - isolate the electrics
  - if your vehicle has a trailer disconnect it
  - call the fire service: cab fires quickly go out of control.

Medical safety

- As an LGV driver you are legally required to be physically fit
- You must notify the DVLA if a physical disability is likely to interfere with your performance
- Take proper rest periods – *fatigue is a killer*
- Acquire a knowledge of basic first aid and carry a first aid kit with you.

Security when delivering

- Don't leave the doors or shutters of your vehicle open while you look for the receiving warehouseman
- Be sure that the person who receives and

signs for the goods is authorised to do so
- Don't leave goods on the pavement if premises are shut
- If your load is valuable don't talk about it. If stopped by police ask to see identification.

Security on the road
- Vary your route and stops – thieves may be studying your routine
- Park your vehicle in sight when you stop, and make sure it is properly locked
- If stopped, even by the police, try to avoid leaving the cab. If your load is valuable, police should escort you to the nearest police station before you leave your cab
- Keep cab doors locked while you are in the vehicle
- If seals are tampered with inform police immediately
- Use proper security parks whenever possible
- Don't leave the vehicle unattended for long periods
- Don't leave the keys in the vehicle – *even for a moment.*

EMERGENCIES – A CHECKLIST

If your vehicle sheds its load you *must act quickly*:
- If possible, park so as to cause minimum disruption to traffic
- Apply handbrake
- Switch engine off
- Make sure that any part of the load still on the vehicle does not slip
- Contact your company immediately and arrange for transhipment or reload
- Move whatever goods you can from the carriageway

- Call the police (and other emergency services if necessary)
- Warn other road users – hazard triangles, hazard warning lights, etc.

15
Coupling and Uncoupling

Demountables
Drawbar Outfits
Articulated Vehicles
Skip Loaders

This chapter explains the procedure for
coupling and uncoupling the four most
popular types of detachable. These are:

- **Demountables**
  (a) rear body lift and roll-off system
  (b) full body lift system

- **Drawbar outfits**
  (a) free-fall drawbar
  (b) pre-set aligning drawbar

- **Articulated vehicles**
  (a) fifth wheel
  (b) automatic

- **Skip loaders**

DEMOUNTABLES

There are many independent demountable
body systems, and their operational
characteristics vary. To ensure safety and
efficiency you should always consult the
manufacturer's instructions. The
information in this section is a *general guide*
to two typical systems.

Rear body lift and roll-off system
DEMOUNTING

Before demounting
- Make sure that the ground where you are working is *flat* and *firm*.
- Select a safe and sensible position for the demounted body *before you start*.

Demounting
- Activate isolation valve (warning light in cab will come on)
- Release all locking points between body and chassis
- Raise the rear end of the body using the vehicle's lifting gear (where the lifting gear is designed to elevate the chassis as well as the body release all locking points *after elevation*)
- Lower landing legs at rear of body and lock securely in place
- Lower the body slowly – the landing legs will take the weight
- Ensure the over-run safety device is in place (if not automatic)
- Drive forward, slowly and in a straight line, until the front of the body is in line with the rear axle of the chassis
- Stop – apply handbrake
- Lower front landing legs (if necessary, elevate again for ground clearance, but don't forget to lower again afterwards)
- Lock landing legs securely in position
- Release the over-run device (if not automatic)
- Drive chassis slowly and in a straight line from the body
- De-activate isolation valve – warning light will go out.

After demounting
- Check the body – is it safe?
- Ensure that all landing legs are securely in place.

Rear body lift and roll-off system
MOUNTING

Before mounting
- Ensure that all landing legs are securely in place
- Check that there are no obstructions to hinder a safe connection
- Ensure that rear of body is clear before mounting, as the body moves back during mounting operation.

Mounting
- Activate isolation valve (warning light in cab will come on)
- Reverse chassis, slowly and in a straight line, until rear axle is in line with front of body
- Stop – apply handbrake
- Raise the front landing legs of body (elevate if necessary), and lock securely in position
- Reverse fully under body
- Raise the rear end of the body – the weight will be removed from the landing legs
- Raise the landing legs and lock securely in position
- Check body and chassis alignment
- Lower body
- Lock all points securing body to chassis
- Ensure that the safety lock is in the *on* position
- De-activate isolation valve (warning light will go out).

After mounting
- Check all locking points
- Check the vehicle – is it safe?

Full body lift system
DEMOUNTING (see Figure 49)

1. Unlock all locking points

2. Elevate body – lower legs

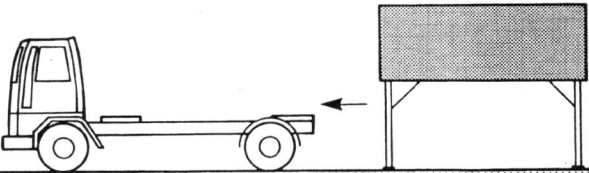

3. Lock legs – move away in line

Figure 49
Full Body Lift System
To dismount container body

Before demounting
- Make sure that the ground where you are working is *flat* and *firm*.
- Select a safe and sensible position for the demounted body *before you start*.

Demounting
- Activate isolation valve (warning light in cab will come on)
- Release all locking points between chassis and body
- Raise the body, using the lifting gear – ensure clearance between landing legs and ground
- Lower all landing legs and lock securely in position
- Lower body slowly – check clearance between body and chassis
- Make sure that the body rests firmly on all legs
- Check clearance – drive away slowly and in a straight line
- De-activate isolation valve (warning light will go out).

After demounting
- Check the body – is it safe?
- Ensure that all landing legs are locked securely in place.

Full body lift system
MOUNTING (see Figure 50)

Before mounting
- Ensure that all landing legs are securely in place
- Check that there are no obstructions to hinder a safe connection.

Mounting
- Activate isolation valve (warning light in cab will come on)
- Reverse fully underneath body
- Stop – apply handbrake

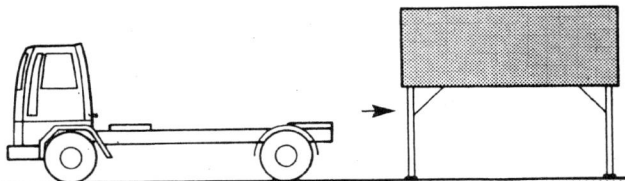

1. Reverse in line

2. Elevate body

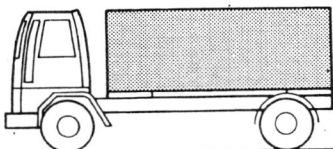

3. Check all locking points

Figure 50
Full Body Lift System
To pick up container body

- Raise the body so that the weight is removed from all landing legs
- Raise landing legs and lock securely in position
- Lower body slowly on to chassis and secure all locking points
- De-activate isolation valve (warning light will go out).

After mounting
- Check all locking points
- Check the vehicle – is it safe?

DRAWBAR OUTFITS

The free-fall drawbar requires a second person to ensure correct alignment. The pre-set aligning drawbar can be operated by the driver alone.

Uncoupling (see Figure 51)

Before uncoupling
- Make sure that the ground where you are working is *flat* and *firm*.
- Select a safe and sensible position for the trailer *before you start*.

Uncoupling
- Apply trailer brake
- Check trailer wheels
- Turn off air taps (if fitted)
- Disconnect air and electrical lines
- Remove safety pin and disconnect drawbar connection
- Put trailer number plate in cab
- Drive away slowly.

After uncoupling
- Check trailer brake
- Check trailer wheel chocks.

1. Park safely – apply the trailer brake

2. Chock trailer wheels – disconnect trailer

3. Move off – check trailer

Figure 51
Uncoupling the Drawbar

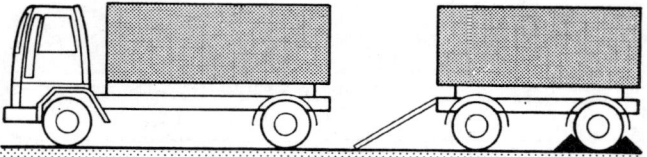

1. Reverse in line with trailer

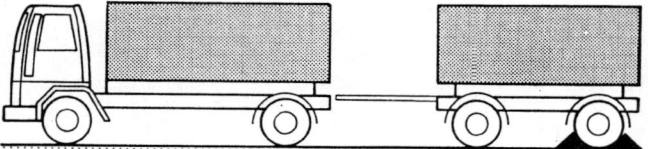

2. Align drawbar with prime mover

3. Connect trailer – remove wheel chocks

Figure 52
Coupling the Drawbar

Coupling (see Figure 52)
Before coupling
- Ensure that trailer parking brake is *on*
- Ensure that wheel chocks are in place
- Ensure that drawbar is at *right angles* to the front of the trailer.

Coupling (two-person operation)

Driver
- Reverse slowly – *know exactly where your assistant is positioned*
- Minimise engine noise – you must be able to hear his instructions

Assistant
- As vehicle is reversed make sure that its coupling box (Figure 53) is in line with the drawbar
- Tell the driver to stop as soon as the coupling connection is made
- Lock the coupling mechanism – and *stand clear*

Driver
- When you are sure that your assistant is clear of the vehicle select a low gear and tug forward to test coupling
- Stop – apply handbrake.

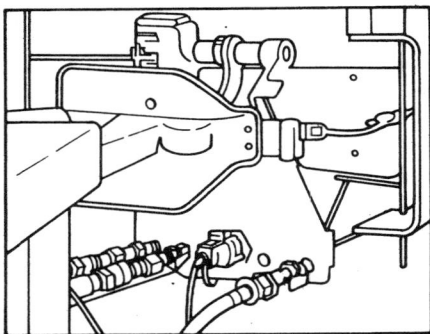

Figure 53
Coupling Box

Assistant	• Connect air and electrical lines
	• Turn on air taps (if fitted)
	• Attach appropriate number plate to rear of trailer
	• Remove wheel chocks
	• Release trailer brake.

After coupling
- Check all tyres and lights (including indicators)
- Test brakes at earliest opportunity.

ARTICULATED VEHICLES

There are two types of coupling system on articulated vehicles. These systems are *not* interchangeable.

Fifth wheel
UNCOUPLING (see Figure 54)

Before uncoupling
- Make sure that the ground where you are working is *flat* and *firm* – if you are in any doubt place a railway sleeper (or similar) underneath the trailer legs
- Select a safe and sensible position for the trailer *before you start*.

Uncoupling
- Apply trailer brake
- Lower trailer legs and return handle to holder
- Turn off air taps (if fitted)
- Disconnect air and electrical lines
- Remove safety catch and disconnect fifth wheel coupling
- Put trailer number plate in cab
- Drive away slowly.

After uncoupling
- Check trailer brake is still *on*

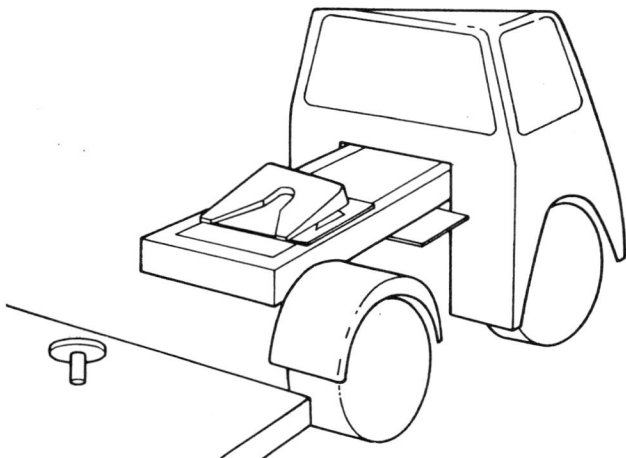

Figure 54
Fifth Wheel Coupling System

- Ensure safety of trailer.

Fifth wheel
COUPLING (see Figure 54)

Before coupling
- Check that trailer brake is *on*
- Ensure that the trailer height is level with the unit

Coupling
- Reverse unit *slowly* under trailer until coupling engages
- Select a low gear and tug forward to test coupling
- Ensure that the king pin is properly engaged and apply safety catch
- Connect air and electrical lines (see Figure 56)
- Turn on air taps (if fitted)
- Raise trailer legs and return handle to holder
- Release trailer brake
- Attach appropriate number plate to rear of trailer.

After coupling
- Check all tyres and lights (including indicators)
- Test brakes as soon as possible.

Automatic
UNCOUPLING (see Figure 57)

Before uncoupling
- Make sure that the ground where you are working is *flat* and *firm*
- Select a safe and sensible position for the trailer *before you start*.

Uncoupling
- Apply trailer brake
- Turn off air taps (if fitted)
- Disconnect air and electrical lines

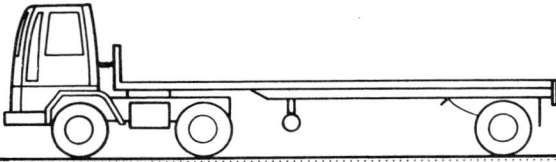

1. Firm ground

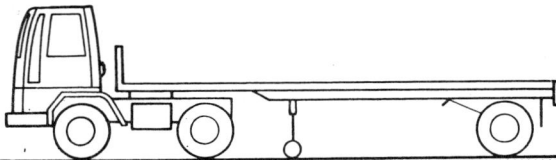

2. Apply trailer brake
   Lower landing gear
   Disconnect air line and electronics

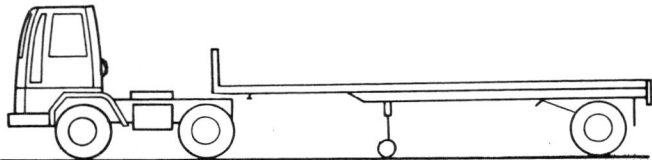

3. Disconnect fifth wheel and move off

Figure 55
Fifth Wheel Coupling

- Put trailer number plate in cab
- Select a low forward gear and release
  handbrake
- Release trailer locking mechanism, and
  drive away *slowly*.

162

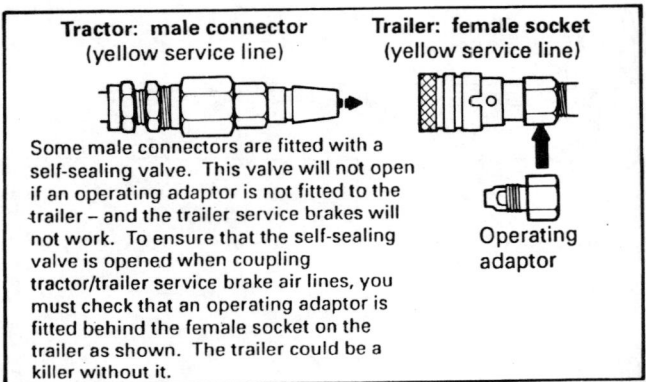

Figure 56
Air Line Connections

After uncoupling
- Check trailer brake is still *on*
- Ensure safety of trailer.

Automatic
COUPLING (see Figure 57)

Before coupling
- Check that trailer brake is *on*
- Check landing gear and turntable combination and make sure that it is at the correct angle for coupling.

Coupling
- Reverse *slowly* under trailer until coupling engages
- Select a low gear and tug forward to test coupling
- Connect air and electrical lines
- Turn on air taps (if fitted)
- Release trailer brake
- Attach appropriate number plate to rear of trailer.

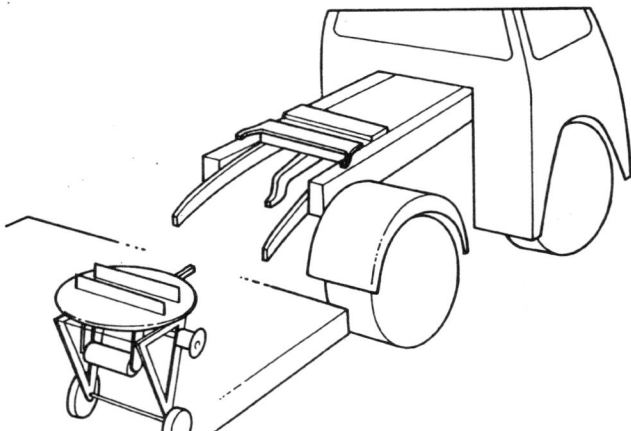

Figure 57
Automatic Coupling System

After coupling
- Check all tyres and lights (including indicators)
- Test brakes as soon as possible.

SKIP LOADERS

Operational characteristics vary according to type or model, and you should always consult the manufacturer's instructions. The information in this section is a *general guide*.

Raising the load (see Figure 59)
- Align truck and container
- Depress clutch and engage PTO (power take-off)
- Lower boom to required height
- Adjust position of vehicle so that the chains are central to the container
- Apply handbrake

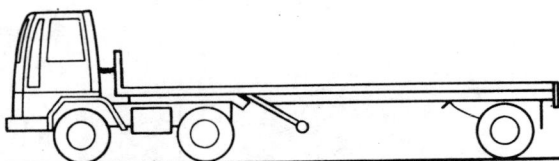

1. Firm ground

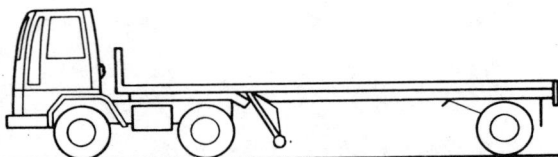

2. Disconnect air and electrical lines
   Release landing gear

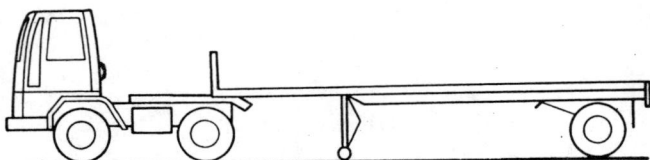

3. Move off

Figure 58
Automatic Coupling

- Attach key plate at the end of the four
  lifting chains to the lugs on the side of the
  container – *make sure that the chains are
  not twisted*

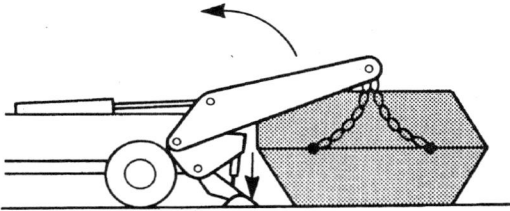

Figure 59
To Raise the Load

- Lower jack legs into position if required – the rear wheels of the vehicle *must not* be lifted off the ground
- Raise the load using the hydraulic controls – use the accelerator to control the speed of the lift
- Hold the control in position until the container is safely in place on the vehicle platform
- Raise the boom slightly to remove the slack from the chains – this is particularly important if you are carrying empty containers
- Raise the jack legs (if necessary)
- Disengage PTO before driving away.

Dumping the load (see Figure 60)
- Make sure that there are no obstructions to the safe dumping of your load
- Apply handbrake
- Depress clutch and engage PTO
- Using hydraulic controls lift the container slightly. (It will engage the trip hook and be tilted over the rear apron when the boom is allowed to travel further)
- When the container reaches an angle steep enough to tip the load stop the boom
- When the container is empty return the boom so that the container settles on the vehicle platform

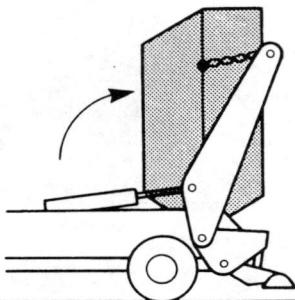

Figure 60
To Dump the Load

- Return control to neutral and disengage PTO.

Note: When dumping heavy loads or materials which do not tip freely lower jack legs to provide support for the over-hung load.

Lowering the load (see Figure 61)
- Ensure that there are no obstructions to safe unloading
- Apply handbrake
- Lower jack legs
- Depress clutch and engage PTO

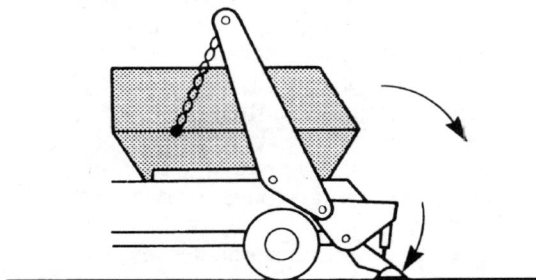

Figure 61
To Lower the Load

- Using the hydraulic controls lower the container to the ground and allow the chains to slacken
- Raise jack legs
- Detach chains from container
- Raise boom
- Return control to neutral and disengage PTO.

Note: These instructions also apply when off-loading *empty* containers.

16
Reversing With Safety

There can be no standard routine for safe reversing, but this chapter offers some basic rules for drivers and guides.

Drivers
- NEVER reverse from a side road into a main road
- ALWAYS check that there are no pedestrians – particularly children – or obstructions behind your vehicle
- NEVER rely on your mirrors alone
- NEVER reverse further than is absolutely necessary
- Whenever possible arrange for an attendant to guide your manoeuvres
- Agree a code of signals with your guide – four hand signals are suggested in Figures 62 to 65.
- Take signals from your guide only, but react to an emergency signal to *stop* from *anyone*.

Guides
- Make sure that you understand the driver's requirements and that you have agreed a code of signals
- *Be flexible* – you may need more signals than the ones suggested
- Position yourself so that you can be clearly seen by the driver, but well out of danger

Figure 62
Stop

Raise the right hand
and arm vertically,
palm facing the driver,
fingers extended but
together and shout
'stop' loud and clear

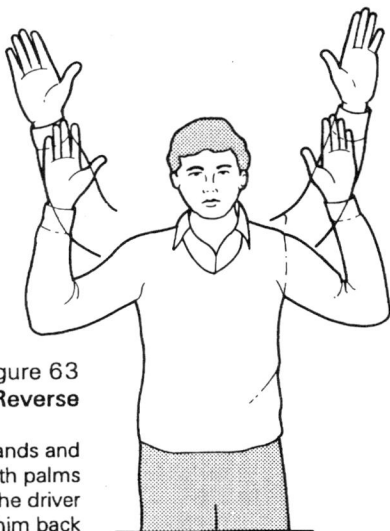

Figure 63
Reverse

Use both hands and
forearms with palms
facing the driver
'pushing' him back
and calling 'go on' if
necessary

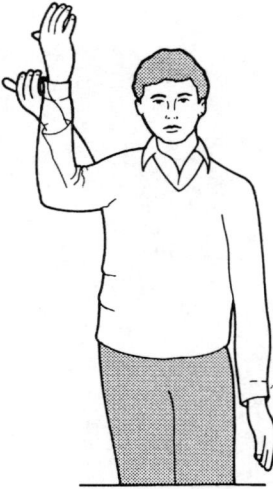

Figure 64
Advance

Use the right hand
and forearm vertically,
palm away from the
driver 'drawing' him
onwards and calling
'come on' if necessary

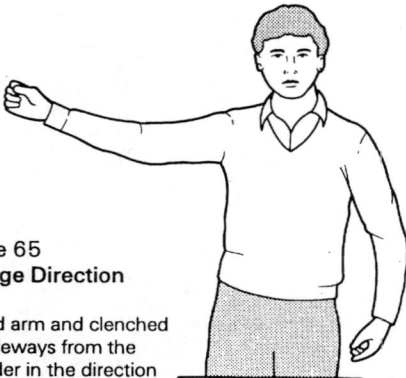

Figure 65
Change Direction

Extend arm and clenched
fist sideways from the
shoulder in the direction
the rear of the vehicle is
to be steered

- Give your signals clearly and in plenty of time for the driver to react
- Whatever the type of vehicle being reversed, the 'change of direction' signal indicates to the driver the direction the *rear* of the vehicle is to be steered.

17
Driving Overseas

Driving
Recovery
Drivers' Accommodation
Drivers' Subsistence
Summary

DRIVING

Vehicles in Europe are driven on the *right*,
but this should present few problems to
British drivers with right-hand drive vehicles.
You should always concentrate and take
care, however, particularly when driving
overseas for the first time.

- Take special care at *road junctions* and
  traffic islands. Traffic approaching from
  the right has priority (except in Germany
  and Sweden)
- Take care when *overtaking*. As you may
  be pulling out 'blind' from behind other
  large vehicles, you should leave extra
  space between yourself and the vehicle in
  front to gain better vision. Additional
  nearside mirrors are very helpful
- Remember that cars on *autobahns* or
  autostradas are likely to be travelling at
  far higher speeds than on British roads.
  Make sure the road is completely clear
  before overtaking
- Most *traffic signs* are international but
  you may be unfamiliar with some
  continental signs. It is a good idea to
  keep a small transfer showing these signs
  in a conspicuous position in the cab.

Motoring organisations will supply these
- Motoring organisations (as well as garages and motoring shops) will also supply the *red warning triangles* that you must display on the road behind a broken-down vehicle to warn approaching traffic
- *On-the-spot fines* are common on the Continent, and can be imposed by the police for even minor infringements of the law. They can be heavy but drivers can decline to pay, and have the case heard in court. However, *this may involve a special journey to Europe*, so in the interests of reducing delays it is usually best and cheapest to pay the fine.

Bans on goods vehicles

Some countries do not allow goods vehicles to use the roads at weekends or on public holidays. Make sure you know the regulations for each country you are visiting since public holidays vary and do not always coincide with those of Great Britain. For example:
- France – movement of goods vehicles restricted on certain days (eg Bastille Day) and over some routes
- West Germany – movement of goods vehicles over 7.4 tonnes (7.5 tons) gross weight prohibited between 07.00 hours on Saturdays and 22.00 hours on Sundays and on public holidays
- Switzerland – movement of goods vehicles prohibited on Sundays, public holidays and during certain hours of the night
- Italy – movement of goods vehicles over 4.9 tonnes (5 tons) laden prohibited on Sundays, public holidays, feast days, and during certain hours of the night. On holiday weekends this restriction is extended to cover certain busy roads.

Italy also requires you to obtain permission *before* taking extra long or wide loads into the country.

Check all the appropriate restrictions before you leave the UK.

RECOVERY

Even the best trucks may break down or become involved in accidents. Whatever the reason for a truck being unable to complete its journey you may need some kind of assistance, be it immediate repair or recovery and repatriation.

A truck broken down in Western Europe presents little difficulty, and recovery from Eastern bloc countries should prove no problem. The Middle East, however, is an area from which recovery is difficult, and the most economical course may well be to dispose of the truck there and to return home by air. In the UK operators either rely on the recovery service provided by the vehicle manufacturer or join one of the organisations that provide a reasonably priced and efficient repair and recovery service – Nationwide Breakdown Recovery Services (Commercial) Ltd or BRS Rescue, for example.

If you are driving a hired truck be sure of the repair and recovery facilities that the hiring company provides in the various countries through which you may be travelling. If you do this you can be sure that the hiring company is represented in those countries.

DRIVERS' ACCOMMODATION

Drivers who travel regularly to the Continent usually find their own suitable accommodation at acceptable prices since there are a great number of cafés, hostels and roadhouses on all major routes and in towns.

British drivers can join the French organisation *Les Routiers*, which vets eating and sleeping places for lorry drivers on the Continent. Approved establishments display a red, white and blue circular *'Les Routiers'* plaque and can be relied on to provide a welcoming atmosphere, food at reasonable prices for good helpings, and generally good value for money. Members receive a guide showing approved places across the Continent and a few in the UK, a badge to display on their vehicle and membership of the UICR (*Union International des Chauffeurs Routiers*), a European lorry drivers' union. Other benefits of membership include insurance protection in the event of accident abroad, benefits to next-of-kin in the event of death following an accident in France, an international membership which is recognised on the Continent as a valid identity document (it has a passport-type photograph of the holder attached), and an international notebook containing useful information for the trans-European driver. It is common practice on the Continent (and a legal requirement in France) for restaurants and cafes to display a priced menu outside so the driver can check what is offered and what the charges are. If a menu is not displayed the place is best avoided.

DRIVERS' SUBSISTENCE

Since the level of expenditure by drivers on subsistence while abroad is normally much greater than when operating in the UK, make sure you get evidence of expenditure to back up expense claims. The Inland Revenue

can ask to see evidence of payments to
drivers for subsistence.

SUMMARY

International trucking is nothing to be
frightened of provided that you:
- carry *all* the correct and necessary
  documentation
- plan your route meticulously
- know the regulations that apply in the
  various countries you will be visiting, and
  abide by them, and
- are fully prepared in every respect.

18
Driver
Communications

CB – Use And Abuse
A Breaker's Glossary
Frequency Map

CB – USE AND ABUSE

The CB situation is now stabilising. After the
initial surge of enthusiasm, brought on by an
overdose of American trucking movies and
the long drawn-out battle for the legalisation
of CB in this country, it would seem that
interest is now declining as rapidly as the
sales of legal CB radio sets. This is
misleading. The brief flirtation with CB by
the media and the general public has faded
out, leaving behind a hard core of people
who have always found CB useful for their
everyday lives, and who probably always
will. Among these, LGV drivers are a
prominent group – some estimates indicate
that as many as 50 per cent of all larger
British goods vehicles are equipped with CB.
The reduction in sales of CB equipment
reflects a stabilisation around this hard core
of working users, rather than a continuous
decline. It will probably prove healthy for
CB that the fad is over.

Many people have been quietly using
some form of CB for a long time, sales reps
operating in large territories, for example.
But very often it is not the long-distance

'juggernaut' drivers so often associated with the image of CB who find it most useful. On the contrary, it is the drivers of short-haul vehicles, such as skip lorries, who gain the greatest working use from their sets. Even if a neighbourhood is reasonably familiar to such a driver, he may well waste a considerable part of his working day searching for a particular street. The driver with CB can just call up help from someone who knows the area well, receiving his directions as he is driving along, and describing his route as he sees it unfolding before him. There are similar advantages for drivers of vehicles travelling in convoy. It will not be necessary to pull off the road whenever one or other has something to communicate. For general and long-distance driving the advantages are twofold. On the one hand, there is the pure information aspect. It puts drivers in contact with motorway service areas, with recovery services in the event of a breakdown, and perhaps with their own depots in similar unfortunate circumstances. It can also be very reassuring, as you are turning on to the M6, say, to know that it is clear all the way to Preston. If the road is blocked there is sure to be someone about on channel 19 (the truckers' channel) who can suggest an alternative route.

There are obviously plenty more useful pieces of information that can be gained in this way, and they hardly need to be spelled out. On the other hand, CB can simply make driving more pleasant. It enables people to contact others of similar (or very different) type, in the course of a working day that might otherwise be dull and lonely. Many drivers have struck up friendships on the air, on regular runs, talking to someone for months before ever actually meeting them face-to-face (an 'eyeball' in CB jargon). Even if drivers are not involved in regular trips up and down the motorways, much of the joy of CB is that it puts them in contact with all sorts of people (using channel 14, to meet other 'breakers'), as they are driving. The conversations may be short, as they may be travelling in opposite directions, but that can be part of the charm. No one knows anything about you except what they can gather from what you say, and if you don't like a conversation, you don't have to take part in it. To be realistic, however, in this country, where the long-distance haul is the exception rather than the rule, it is definitely the working value of CB that makes it a good proposition for the LGV driver rather than its more social qualities. What about the cost? At present, only 27MHZ FM is legalised, and this has caused outrage among the thousands of CB users who have been operating AM sets for years. As a result, a lot of people are still using illegal sets, and there does not seem to be much the Government can do about it. A reasonable legal set, with an SWR metre for tuning the aerial, will cost from £20, with no extras, to £70. Illegal sets are currently available for about £35, but be careful here – they could be poor quality sets, or foreign sets using a different frequency that will be no good at all in Britain. The best bet now is to buy a legal FM set from a reputable dealer, whatever the pros and cons of AM may be.

Most LGV drivers use twin aerials. These give a reception band maybe 20 miles ahead and behind, but only a couple of miles to either side. Although this is ideal for motorway driving, a single aerial, giving a circular reception area of shorter range, is quite adequate for local work. Most regular, working CB users insist that their sets have paid for themselves in saved time and fuel within a few months of installation. The

advantages of increased pleasure in driving
and decreased stress are immeasurable.

A BREAKER'S GLOSSARY

A bit 10-1 – Weak or fading out
Ace – CBer with a rather high opinion of
    himself
Appliance operator – A CBer who doesn't
    know anything about his set
Auntie Beep – The BBC

Back Door – Vehicle behind you watching
    what's on the motorway
**Back side, back stroke, bounce around, flip-
    flop** – Return trip
Back of the hammer – Slow down
Bean store – Restaurant or road stop where
    food is served
Bear – Police
Bear bait – Someone driving without a radio
Bear den – Police station
Bear's lair – Police station
Bear bite – Speeding ticket
Bells – Hours; four bells is four o'clock
Bending my ears – Bidding me
Bending my windows – Good copy
Better half – Wife or husband
Big brother – Home Office, GPO
Big circle – North Circular Road in London
Big dummy – Truck driver term
Big wheels – Lorry
Big slab – Motorway
Big Smoke – London
Bird cage – Heathrow Airport
Black box – CB radio
Bleeper breaker – Transmitting a bleep when
    you let off your mike, signifies end of
    transmission
Blue light – Police vehicle
Bottle shop – Pub
Bottom of the shop – Channel 1

Break – Call a station
Break, break; Breakity, break – Breaker
Break, etc – Signal that you want to get on a channel
Breaker – CB user
Bring it back – Said at end of transmission
Brown bottles – Beer
Brown bottle shop – Pub
Buzby – GPO

Candles – Years
Chalk a block – All channels filled up with breakers
Channel master – Breaker who tries to monopolise or control a channel
Chicken coop – Truck inspection station
Coming out the windows – Good copy
Cop shop – Police station
Copy? – Did you hear me?
Copying the mail – Listening to other breakers
Cubs – Police
C'mon – Transmit

Dinosaur juice – Petrol
Dodgy – Risky
Don't feed the bears – Don't get arrested
Doughnut – Roundabout
Draggin' wagon – Wrecker

Ear wigging – Monitoring, listening on the side to an ongoing conversation
Easy chair – Sitting in the middle of a convoy, also called the rocking chair
Eights, Eighty-eights – Best wishes
Eleven metres – CB band
Eyeties – Italian stations

Factory – Place of work, depot
Ferry lights – Traffic lights
Fetch it back – Said at end of transmission
Feed the bears – Pay a ticket
First person – Your name
Fluff stuff – Snow

Front door – Vehicle ahead of you, or at the
head of a convoy

Gear Jammer – Truck driver
Go juice – Petrol
Going breaker break – Leaving the air
Going down – Signing off the air
Good lady – Female CBer
Good numbers, goodly numbers – Best wishes
Green light – Clear road on up ahead
Green shield stamps – Money
Ground clouds – Fog

Handle – CB nickname
Hammer – Accelerator pedal
Harry Rags – Cigarettes
Holding on to your mudflaps – Driving close
behind you

In a short – Shortly
In a short short – Very shortly

Jam sandwich – Police vehicle, white with red
stripe through the middle
Jaw jacking – Long-winded conversation

Keep it clean and don't be seen – Don't get
arrested
Klicks – Kilometres
Keep 'em between the ditches – Safe driving!

Land line – Telephone
Little wheels – Car or other four-wheeled
vehicle

Motion lotion – Petrol
Mud – Coffee

Negatory – No
Nelly Kelly – TV
Noddies – Motorcycle police
Nosebag – A meal

On channel – On the air
On the peg – Legal speed limit

One armed bandit – Petrol pump
Over your shoulder – Behind you

Pedal to the metal, hammer down – Speed up
Peeling off – Leaving the motorway
Personal – Your name
Plain wrapper – An unmarked police car
Pushing big wheels – Driving a lorry
Pushing wheels – Driving a vehicle

Ratcheting – Talking
Reading the mail – Listening to the channel
Roller skate – A small car

Smokey, smokey the bear – Police
Smokey report – Location of police
Smokey on rubber – Police moving
Smokey town – London
Spaghetti junction – Birmingham
Sparks – Radio technicians
Square wheels – Stationary vehicle
Station on the side – CBer wanting to join
 conversation
Suicide jockey – Lorry driver carrying a
 hazardous load
Super slab – Motorway

Taking pictures – Police using radar
Ten-ten till we do it again – Sign off
Tower Town – Blackpool

Walking all over you – A louder station is
 drowning your signal
Wall to all and tree top tall – Loud and clear
Wally – Silly CB user
Watering hole – Pub
Wrapper – Vehicle

XYL – Wife

YL – Young lady

Z's – Sleep
Zoo – Police headquarters

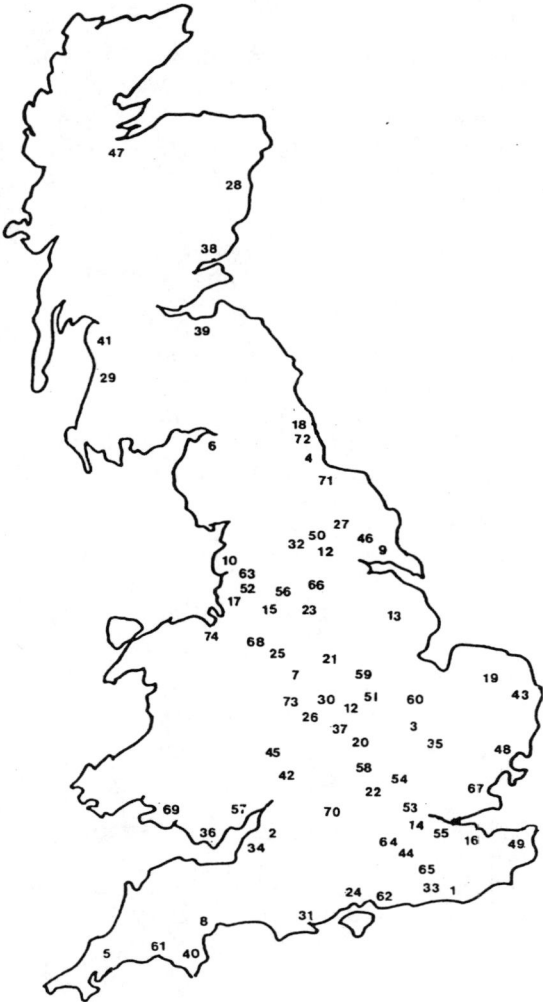

Figure 66
Frequency Map

BBC Local Radio Stations
(Main frequencies only)

Radio Station	MW kHz	FM MHz
1. Radio Sussex and Surrey	1161	104.5
2. Radio Bristol	1548	95.5
3. Radio Cambs	1026	96.0
4. Radio Cleveland		95.0
5. Radio Cornwall	630	103.9
6. Radio Cumbria	756	95.6
7. Radio Derby	1116	104.5
8. Radio Devon	55	103.4
9. Radio Humberside	1485	95.9
10. Radio Lancashire	855	103.95
11. Radio Leeds	774	92.4
12. Radio Leicester	837	104.9
13. BBC Radio Lincs	1368	94.9
14. GLR (London)		94.9
15. GMR (Manchester)		95.1
16. BBC Kent	774	96.7
17. Radio Merseyside	1485	95.8
18. Radio Newcastle	1458	95.4
19. Radio Norfolk	855	95.1
20. Radio Northampton		104.2
21. Radio Nottingham	1584	103.8
22. Radio Oxford		95.2
23. Radio Sheffield	1035	104.1
24. Radio Solent	999	96.1
25. Radio Stoke	1503	94.6
26. Radio WM	1458	95.6
27. Radio York	666	103.7

For a booklet containing a more detailed frequency listing contact: BBC Engineering Information, White City, 201 Wood Lane, London W12 7TS Tel: 0345 010 313. In addition, the following BBC Local Radio Stations should be added to the list above: BBC CWR (Warwick area) FM 94.8; BBC Dorset FM 103.8; BBC Essex FM 103.5, MW 765; Radio Gloucs FM 104.7; Radio Guernsey FM 93.2, MW 1116: BBC Hereford and Worcester FM 94.7 (Hereford), FM 104.0, MW 738 (Worcester); Radio Jersey FM 88.8, MW 1026; Radio Shropshire FM 96.0; Radio Berkshire FM 104.1; Radio Suffolk FM 103.9; BBC Three Counties Radio (Luton and Milton Keynes area) FM 95.5, MW 1161 and BBC Wiltshire Sound FM 103.50.

Independent Local Radio Stations*

Radio Station	MW kHz	FM MHz
28. Aberdeen: NorthSound	1035	96.9, 103.0
29. Ayr: West Sound Radio PLC	1035	96.7, 97.5
30. Birmingham: 96.4 FM BRMB		96.4
31. Bournemouth:		
2CR Classic Gold	828	
2 CR FM		102.3
32. Bradford: The Pulse		97.5
33. Brighton: South Coast Radio	1323	
34. Bristol: Galaxy Radio		97.2, - 101.0
35. Bury St Edmunds: SGR-FM	1251	96.4
36. Cardiff: Touch AM	1359	
37. Coventry: Mercia		97.0, 102.9
38. Dundee/Perth: Radio Tay		
Dundee area	1161	102.8
Perth area	1584	96.4
39. Edinburgh: Forth FM		97.3/ 97.6
40. Exeter/Torbay: Devon Air Radio		
Exeter area	666	97.0
Torbay area	954	96.4
East Devon		103.0
41. Glasgow: Clyde 1		102.5
Clyde 2	1152	
42. Gloucester & Cheltenham:		
Severn Sound Super Gold	774	
Severn Sound		102.4 & 103.0
43. Great Yarmouth & Norwich: Radio Broadland	1152	102.4
44. Guildford: Radio Mercury FM (West)		96.4
45. Hereford/Worcester Radio Wyvern		
Hereford area	954	97.6
Worcester area	1530	102.8
46. Humberside: Viking FM		96.9
47. Inverness: Moray Firth Radio	1107	7.4
48. Ipswich: SGR-FM	1170	97.1
49. East Kent:Invicta/ Supergold	603	
50. Leeds: Radio Aire FM		96.3
51. Leicester: Leicester Sound		103.2
52. Liverpool: City FM		96.7
53. London:		
Capital FM (General)		95.8
London Newstalk Radio	1152 AM	
54. Luton/Bedford: Chiltern Radio		
Luton area		97.6
Chiltern Radio (East) Bedford Area		96.9
55. Maidstone & Medway: Invicta FM		103.1
56. Manchester: Piccadilly Gold	1152	
57. Newport (Gwent): Touch AM/	1305	
58. Northampton: Northants Radio		96.6
59. Nottingham: Trent FM		96.2/ 96.5
60. Peterborough: Hereward Radio 102.7 FM		102.7
61. Plymouth: Plymouth Sound	1152	97.0/ 96.6
62. Portsmouth: South Coast Radio	1170 & 1557	
63. Preston & Blackpool: Red Rose Gold	999	
64. Reading: 210 Classic Gold Radio	1431	
65. Reigate: Radio Mercury FM (East)		102.7

66. Sheffield & Rotherham:		
Hallam FM		
Sheffield Area		97.4
Rotherham area		96.1
67. Southend/Chelmsford:		
Essex Radio		
Southend Area		96.3
Chelmsford Area		102.6
68. Stoke-On-Trent:		
Signal Gold	1170	
69. Swansea: Swansea Sound	1170	96.4
70. Swindon:		
Brunel Classic Gold		
Swindon Area	1161	
West Wiltshire	936	
71. Teeside: Great North Radio	1170	
72. Tyne and Wear:		
Great North Radio	1152	
73. Wolverhampton &		
Black Country: Beacon		
Radio		97.2
		& 103.1
74. Wrexham & Deeside:		
Marcher Coast 96.3 FM		96.3

*The range of independent local radio stations on
offer is vast. The list above offers a selection. It
should not be used as a complete guide.

19
First Aid

Danger
The Accident
What's Wrong?
Breathing (Respiration)
Bleeding
Shock
Unconsciousness
The Recovery Position

Truck drivers spend most of their working day on the road, so it is likely that they will, at some stage, be first to the scene of a motor accident. In such circumstances knowledge of basic first aid could keep victims alive until skilled help arrives. It must be emphasised that the information in this chapter is *only a general guide*.

Although the text has been approved, it is no substitute for a recognised first aid course.

When you arrive at the scene of an accident be clear about the priorities.

DANGER

Your first task is to eliminate danger, either by moving the danger from the casualty, or by moving the casualty from the danger. A pile-up or a fire are the two great threats.

Moving the danger from the casualty
- Organise bystanders to slow down and divert the traffic
- Use hazard warning lights on strategically placed vehicles

- Switch off engine and apply parking
  brakes of vehicles involved in accident
- Ensure that no one in the area smokes
- Look out for petrol leaks and other
  sources of danger.

When you are satisfied that casualties are in
no further danger call the emergency services
(telephone 999), and give the essential
information clearly:
- Exact location of accident
- Number of people involved
- Types of vehicle involved.

It is always wise to send a car in each
direction with this information. If there is
any doubt about the services needed,
telephone the police; they will notify those
additional services that are required.

Moving the casualty from the danger
This should only be undertaken in *extreme
circumstances*. The victim of an accident
should not be moved unless his life is
seriously threatened by the danger.

Note: If it is necessary to move an injured
person from the roadway, note his exact
position beforehand. The police may require
this information.

THE ACCIDENT

As a first aider your main concern at an
accident is making sure it is safe to attend to
the casualty. This is done by assessing the
situation; what can be seen or heard? In the
case of an accident involving noxious
substances you should also be aware of
smell. Only if you are sure it is safe should
you attempt to give first aid to the casualty.
In cases where it is obvious that a person is
trapped, or where a major accident has

occurred, do not delay in calling the emergency services. At this stage bystanders can be used to summon help and bring the first aid kit, leaving the first aider to attend to the casualty.

Message to the emergency service

In cases of major accidents or sudden illness, the casualty's best chance of making a full recovery is by getting skilled medical aid to them as soon as possible. It is important that the first aider knows the correct method of obtaining the emergency services. It should be remembered that in many circumstances the person calling for help may have to go through the switchboard before they can dial the 999 number.

Having got through to the operator they will be asked which service they require. The ambulance operator is trained to obtain certain information from you. The telephone receiver should not be replaced until the operator tells you to do so.

WHAT'S WRONG?

Before you can treat a casualty you have to find out what is wrong. You do this by using three things:
- Where?
- When?
- How did it happen?

The casualty is the best person to give you this information. If this is not possible, a bystander.

Symptoms

What are the symptoms the casualty feels? They can only be obtained from the casualty, and may include:
- Pain

- Thirst
- Nausea
- Dizziness
- Hot or cold.

Signs

These are what a first aider may see, hear, feel or smell. These may include:
- Objects lying around
- Tools or any sort of machinery
- Bottles that may contain tablets or containers of hazardous substances.

Signs are also things you can find on the casualty, these can include:
- Pulse
- Colour
- Bleeding
- Broken bones
- Odour on breath
- Reaction to pain

Using the history, signs and symptoms, the first aider will be helped to make a diagnosis.

BREATHING (RESPIRATION)

Oxygen is essential for us to maintain life; the way we obtain this is by breathing. To

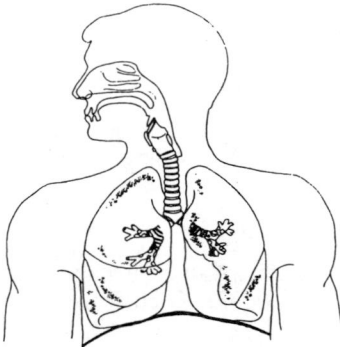

Figure 67

breathe the brain sends messages to the chest muscles and the diaphragm which contract, expanding the chest. This draws air into the mouth and nose, down the windpipe and into the lungs.

When the muscles relax, the chest falls and carbon dioxide and other waste products are breathed out.

Asphyxia
If a person stops breathing the condition is known as asphyxia, which means a lack of oxygen to the brain and other vital organs. This results in the death of nerve cells in the brain, meaning the heart and lungs cease to function effectively.

There are many reasons for this happening, they include:
- Things that can affect the airway and lungs:
  — Choking
  — Hanging
  — Obstruction
  — Smothering
- Things that affect the brain and nerves:
  — Head injuries
  — Spinal cord injuries
  — Certain medical conditions
- Things that affect the air we breathe:
  — Cyanide
  — Carbon dioxide
  — A decrease in the oxygen supply.

Signs and symptoms
The face becomes congested with blueness around the lips and ear lobes. Breathing becomes difficult, noisy and froth may appear around the mouth. If the condition is not treated quickly breathing will stop.

Treatment
Remember you cannot resuscitate a casualty if the cause is still present. The following

guidelines form a simple system by which a
first aider can keep a casualty's brain
supplied with oxygenated blood until skilled
medical aid arrives.

- **Assess** safety – yourself, casualty and
  others.
- **Speak** shout and shake – Do they
  respond?
- **Help** – Get somebody to phone for an
  ambulance immediately.

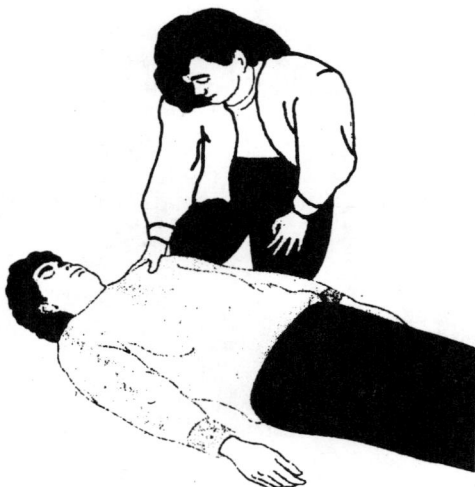

Figure 68
Assessing the Situation

A.   Open airway by lifting casualty's chin

B. Check breathing by placing your ear next to the casualty's mouth to feel the breath.

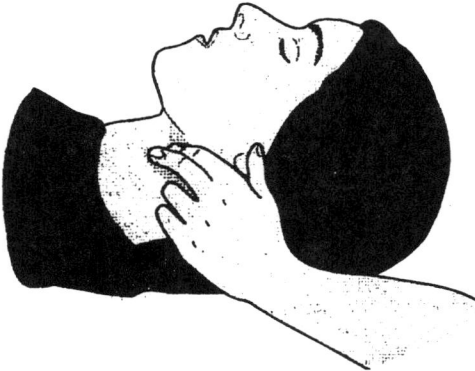

C. Check circulation by finding the person's pulse. This can be felt in the groove between the casualty's Adam's apple and side of their neck.

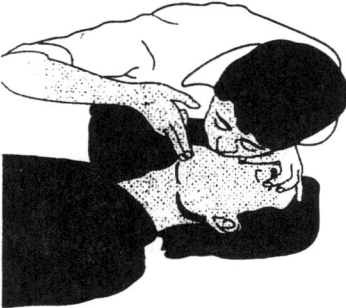

- **Resuscitate the lungs.** If there is a pulse felt, but no breathing, and help is on the way, pinch the casualty's nose and seal your lips around the casualty's mouth. Blow until the casualty's chest wall rises. You should blow for two seconds and then lift and turn your head to see the chest fall. Count for four seconds then repeat this at the rate of ten breaths per minute.

- **Pulse.** If there is no pulse give two breaths as above and continue down the list.

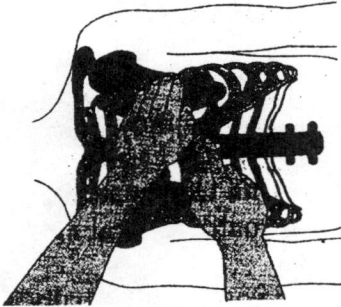

- **Position.** To find the correct chest position bring the two fingers of your hand furthest away from the casualty's head up the rib cage until you feel the point where the ribs meet. Place your two fingers on the breast bone and keep them there. Bringing the heel of your other hand alongside your two fingers, place your original hand on top of the hand closest to the casualty's head, then interlock your fingers. Kneeling as close to the casualty's side as you can, lock your elbows to keep them straight.

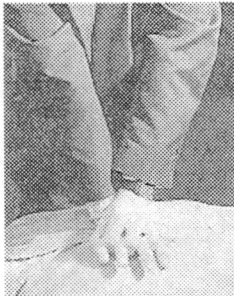

- **Resuscitate the heart.** With elbows locked, so that the force goes straight down, compress the chest to a depth of about two inches. Relax

the pressure and then repeat 14 times. You
should aim to get 15 compressions plus two
inflations in 15 seconds.

- **Evaluate breathing and circulation after two
  minutes** (eight cycles of 15 compressions plus
  two breaths).
- **Start again.** If no breathing or pulse is found
  start your cycle again; from resuscitating the
  lungs.

If you follow these guidelines, you will
provide the casualty's own life support
system with oxygenated blood to enable
survival until medical aid arrives.

If you are alone when finding a casualty,
and on checking ABC find no breathing but
a pulse, give ten breaths then go for help. On
returning to the casualty check ABC again
and action on your findings.

If there is no breathing and no pulse *do
not start* CPR, telephone for an ambulance
immediately. On returning to the casualty,
check ABC and act on your findings. These
two rules apply only where there is a
telephone within five minutes.

Two person CPR
When there are two people performing CPR
the rate is five compressions to one inflation.

All resuscitation should be continued
unless:

- the casualty's pulse returns and they start
  to breathe unaided
- medical aid arrives
- the first aider is no longer able to
  continue.

BLEEDING

Bleeding can be classed as internal or
external. To suffer both can be serious. If an
artery is cut the blood will spurt in time with
the heartbeat, a severed capillary will ooze,
while blood from a vein will gush.

The following steps should be followed when treating external bleeding:

- **Position the casualty.** The position will vary according to the site of the injury. If the bleeding is severe, lay the casualty down and elevate the feet.
- **Expose the wound.** It is necessary to expose the wound so that you can assess the extent of the injury and also ensure no foreign objects are in the wound.
- **Elevate.** Elevating a wound above the level of the heart will slow the bleeding.
- **Pressure.** Pressure should be applied directly onto the wound, or in the case of a wound with a foreign object sticking in or out of it, either side.

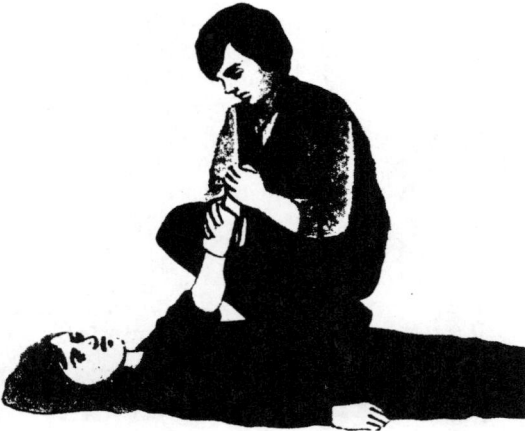

Figure 69
Treatment for Bleeding

In many cases the casualty will be able to control their own bleeding, by applying pressure to, or around, their wound. *Where possible, avoid coming into contact with the blood.*

The measures detailed above may have to be taken until first aid equipment is obtained.

In accordance with HSE guidance, a first aid box in the workplace will contain sterile dressings of various sizes. The purpose of a dressing is to control bleeding, absorb any discharge and to help in the prevention of infection. A first aider should attempt to cover a wound as quickly as possible, taking care not to touch the sterile pad or to breathe or cough over it. When applying a dressing, care should be taken that it is not applied too tightly.

The following signs will be seen if the dressing is too tight.

● Extremities become cold and pale
● There may be an absence of a pulse in a limb below the site of the dressing.

If this occurs the dressing must be loosened immediately.

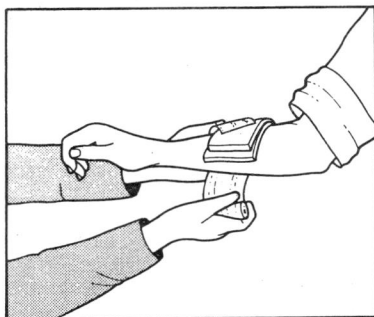

Figure 70
Dressing a Wound

If one dressing fails to stop the bleeding *do not* remove it; apply a further dressing. You can apply a maximum of three dressings. If this fails to control the bleeding, indirect pressure – when an artery is compressed against bone – may have to be applied. This should be attempted when three dressings have failed to stop bleeding.

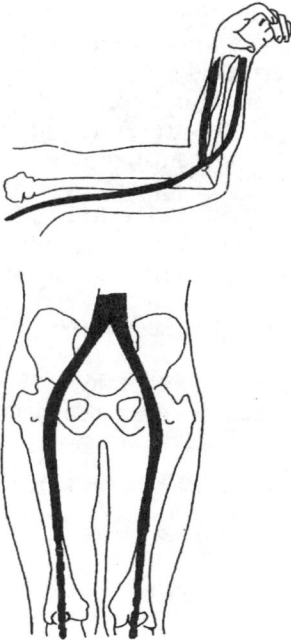

Figure 7 1
Vital Arteries

Pressure should be applied to the artery at
the top of the arm or in the groin. Pressure
should never be applied for longer than 15
minutes. If an ambulance has not arrived
within this time scale, pressure should be
released to allow the blood to flow back into
the fingers or toes. This can be seen by a
return of colour to those parts. If the
bleeding continues, pressure may be applied
for a further 15 minutes.

Wounds with embedded objects
In the case of an object sticking out of a
wound, pressure should be applied to either
side of the injured area. Dressings should be
placed around the wound, helping to control

the bleeding and to protect the injury by
ensuring no pressure is put on the protruding
object. No attempt should be made to
remove the object, as this may cause more
damage, increase bleeding and allow
infection into the wound.

Bleeding from a special area: the scalp
Bleeding from the scalp may be profuse and
a first aider should examine the wound
carefully before attempting to stem the
bleeding. If a fracture is suspected no
pressure should be applied. If a fracture is
not suspected a dressing should be applied
and held in place with a triangular bandage.
If the casualty is conscious, their head and
shoulders should be raised above the level of
the heart.

Penetrating chest wounds
This is a very serious condition which
requires urgent action by the first aider. With
this injury air can directly enter the chest
cavity causing severe breathing difficulties.
The aim of the treatment is to stop air
entering. This may be done by placing your
hand over the wound until the first aid
equipment can be brought to the scene.
When it arrives a thick pad should be placed
over the wound and bandaged firmly in place
making the wound airtight (a clean plastic
bag may also be used to help make the
dressing airtight).
    The casualty should be placed with head
and shoulders raised, leaning towards the
injured side. This will help the good lung to
function, while also preventing blood that is
coughed up draining back into the lungs.

Wounds to the stomach
If not already doing so, lay the casualty
down and keep head and shoulders raised.
Place a coat or something bulky under the

knees, this helps to stop the abdominal muscles stretching and the wound from gaping. A large pad should be applied firmly over the wound, but do not use excessive pressure. If internal organs protrude, do not attempt to push them back into the wound, build up pads around the protruding organs and apply a dressing over the top without touching or putting pressure on the exposed organs.

Internal bleeding
In many cases of internal bleeding there will be no visible signs, but as bleeding increases the signs and symptoms of shock will become more apparent (see Shock). It is important that the first aider obtains a history of illness, or a recent injury.

In some cases bleeding may become visible where the part of the body that is injured has a passage leading to one of the body's orifices. Listed below are some examples.

- Blood mixed with straw coloured fluid coming from the ear or nose may indicate bleeding from the base of a fractured skull
- Blood that is coughed up and is bright red and frothy indicates bleeding from the lungs
- Vomit which contains what appears to be coffee grounds, indicates bleeding into the stomach
- Blood which is visible in a casualty's urine or faeces can provide a first aider with vital clues about the casualty's history.

The treatment for a casualty suffering from internal bleeding is to lay the casualty down and slightly raise the legs. If blood is coming from an ear, turn the head to the side the discharge is coming from to enable it to drain away. If blood is coughed up or vomited, adjust the position by raising the

casualty's head and shoulders. All internal
bleeding requires urgent medical treatment.

SHOCK

Shock is a lack of oxygenated blood to the
brain and other vital organs, which slows
down the functions of the body. Any injury
or sudden illness may cause shock. The
severity of the shock is dependent on the
seriousness and extent of the injury or
sudden illness. Shock can vary from a feeling
of faintness to death. Although shock is
present to some degree in all injuries, it is
clearly associated with the following:
- **A** – Acute heart attacks
- **L** – Loss of plasma (burns and crush
  injuries)
- **L** – Loss of body fluid (diarrhoea,
  vomiting, heat exhaustion)
- **A** – Acute abdominal emergencies
- **C** – Circulation loss (bleeding)
- **E** – Excessive pain.

Signs and symptoms of shock vary
considerably; they may not all be present,
and depend on the cause. The first aider
should look for some or all of the following:
- Pale face
- Cold clammy skin
- Profuse sweating on head and back of
  hands
- Shallow breathing (dependent on cause)
- Rapid pulse (as pulse rate increases so
  will shock)
- Thirst
- Dizziness
- Feeling sick and/or vomiting.

Treatment for shock is dependent on the
injuries. Severe injuries must be treated first
as this will lessen the severity of the shock. If
possible lay the casualty down, loosen

clothing and raise legs (as with bleeding). In addition, keep the casualty warm, reassure them and monitor their pulse and breathing. *Do not move unless in danger. Do not give anything by mouth.* Avoid the latter as this may delay the giving of an anaesthetic. It may also cause the casualty to vomit, so increasing the amount of shock, as more fluid is lost.

UNCONSCIOUSNESS

Unconsciousness is a disruption of the normal working of the brain, and may result from an injury or a medical condition. A person's level of consciousness may change as the first aider is waiting for the ambulance to arrive.

The levels of 'responsiveness' are:
- **A**lert – fully responsive to questioning
- **V**oice – is not fully alert but responds to being spoken to
- **P**ain – does not respond to being spoken to but responds to pain
- Unresponsive – does not respond at all.

Management of the unconscious casualty
The greatest danger to the life of an unconscious casualty is that their airway may become blocked, resulting in asphyxia:
1. On approaching a casualty, ensure it is safe for you to do so. If there is any danger *do not attempt a rescue*, leave it to the professionals.
2. *If* it is safe to do so, the first aider should find out if the casualty is conscious by talking to them and gently shaking their shoulders, unless the history suggests a spinal injury may have occurred. In this case, steady the casualty's forehead by placing your hand on it and then gently shake a shoulder.

The unconscious casualty may still be able to hear and will sometimes react to being spoken to. Remember this: comments you would not make to a conscious casualty should not be made to an unconscious one.

3. If alone shout for help. It is better for you to get someone else to phone for an ambulance, as it is not advisable to leave an unconscious casualty alone.
4. Check: **A**irway; **B**reathing; **C**irculation.
5. Perform a visual check for bleeding. If serious bleeding is noted treat it immediately.
6. A systematic examination of the casualty should then be carried out. The first aider should start at the head and work down, carefully feeling along each limb and comparing each side to the other. When feeling the body, firm but gentle pressure should be used. If an irregularity is found it should be examined, as this may indicate a fracture.

Other indicators to look for are; medic alert and an SOS talisman worn either round the neck or wrist. These may be helpful signs to the casualty's condition. This information can be passed on to the ambulance crew.

THE RECOVERY POSITION

1. Ensure the airway is open. Turn the casualty's head towards you.
2. Place the nearest arm straight above the casualty's head.
3. Push and bend the furthest knee from you away, tucking this leg's foot under the nearer knee.
4. With your hand nearest the head, grip the thumb of the furthest hand and bring this up to protect the casualty's head.

5. Grip the furthest knee from you and bring it upright.
6. Pull down on the knee. The casualty's body will turn over.
7. Place the casualty's free hand under their head, and check the airway is open.

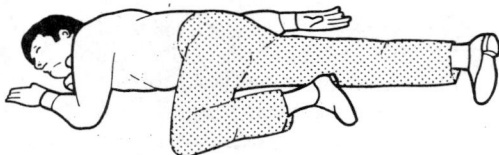

Figure 72
The Recovery Position

The advice offered in this section is no substitute for proper training. All drivers should be encouraged to undertake a formal course in First Aid.

For further information, contact:
Kays Medical
3/7 Shaw Street
Liverpool L6 1HH
Tel: (0151) 207 5117
Fax: (0151) 207 3384

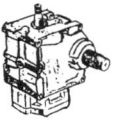
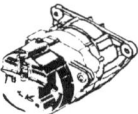

20
Vehicle Faults and Remedies

Questions And Answers

Although drivers of heavy goods vehicles are not expected to be mechanics, they are expected to have a basic understanding of the components which affect the safe handling of their vehicle. The following sample questions and answers will help to give a more comprehensive knowledge of the workings of a large goods vehicle.

QUESTIONS AND ANSWERS

What checks would you make on your vehicle before commencing a journey?

Oil, fuel, water, lights, brakes, steering, tyres, check for air, hydraulic leaks and all warning devices, windscreen, mirrors, spare wheel and carrier, wheel nuts.

What low pressure warning devices are fitted to vehicles?

Gauge, buzzer, light.

What action would you take if your low pressure warning device started to operate while the vehicle was in motion?

Stop, park safely until trouble is rectified.

If you need to pump the brake pedal what has happened?

Shoes out of adjustment, or air in the system.

How do air brakes work?

Air is compressed by an engine-driven compressor and fed to one or more storage tanks. When the brake pedal is depressed a valve opens. This allows the compressed air to move pistons mounted near the wheels which operate the brake shoes.

What is a simple test for locating leaks in the air brake system?

With pressure built up, get someone to depress the brake pedal. A loud hiss will enable you to locate the leak.

With air brakes, why is it dangerous to coast downhill?

Air pressure may not build up when the engine is just ticking over, particularly with a worn compressor.

What is meant by brake fade?

The frictional characteristics of the lining are affected by heat, the brake drums get hot and expand from the shoes, resulting in reduced braking efficiency, especially on long down gradients when the brakes have been excessively used and have become very hot.

In frosty weather, what could prevent air pressure building up in the brake reservoir?

Moisture in the air drawn in by the compressor freezing in the brake system.

What types of glass are used in windscreens?

Toughened and laminated.

How would toughened glass differ from laminated glass in the event of a broken windscreen?

Vision is seriously reduced when toughened glass breaks: this does not happen with laminated.

What is a two-speed axle?

An alternative reduction gear fitted to the back axle which doubles the number of available gear ratios.

If you break down by the side of the road, what must you do?

First warn traffic behind (hazard warning lights) and then diagnose the fault.

If excessive smoke comes from the vehicle's exhaust, what does this tell you and what should you do?

Engine faulty. You must stop your vehicle.

If you break down by the side of the road, how far back should you place a warning triangle?

50–150 metres on the hard shoulder of a motorway.

What action should be taken if the engine temperature gauge indicates overheating?

Stop. Allow the engine to cool down and top up the radiator or header tank with warm water.

A vehicle has just filled up after running out of fuel but will not start. What is the most likely problem?

Air in the fuel system, which requires bleeding from the fuel pump.

21
Vehicle Types and Definitions

Maximum Vehicle Weights
Maximum Vehicle Lengths

Below is a selection of various types of vehicle
showing the numerous wheel arrangement
combinations, along with some of the many
body types used in road haulage operations.

Articulated vehicles
a) Open flat semi-trailer   Single axle
b) Large box van trailer   Tandem axle
c) 6-wheel tractive unit
   (4 steering wheels) (Chinese six)
   with a tandem axle tanker body
d) 6-wheel tractive unit (4 driving
   wheels) with a tri-axle tipper body
e) Double bottom vehicle (an
   articulated outfit drawing a
   second semi-trailer) *not generally
   permitted in this country*.

Rigid vehicles   *y*   *z*
a) Open flat   4-wheeler 4×2
b) Box van   6-wheeler 6×4
c) Luton body   4-wheeler 4×2
d) Tipper vehicle   8-wheeler 4×4
e) Pantechnicon   4-wheeler 4×2
f) Tanker vehicle   6-wheeler 6×2
g) Vehicle and trailer (with a 6-wheeled
   trailer)
h) Vehicle with close coupled four trailer

Note; y = number of wheels on vehicle,
z = number of *driving* wheels

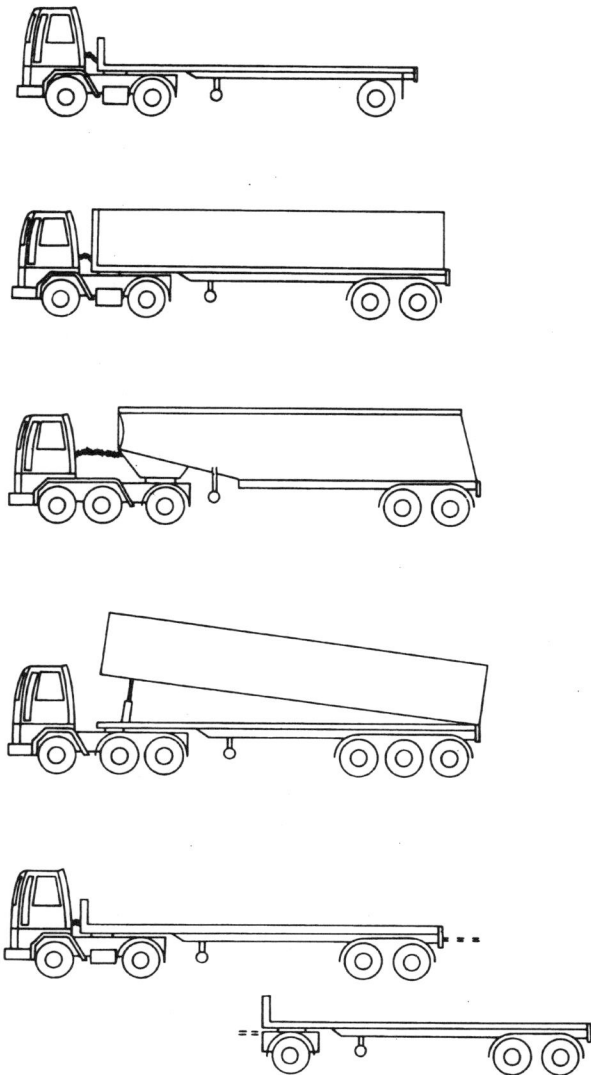

Figure 73
Articulated Vehicles

MAXIMUM VEHICLE WEIGHTS

General Requirements
- Vehicles and trailers must be fitted with a manufacturer's plate.
- Vehicles must have 50% service and 25% secondary braking efficiency.
- Trailers must be drawn by vehicles having 50% service and 25% secondary braking efficiency.
- Road friendly suspension is defined as an air suspension or a suspension equivalent to an air suspension. Air suspension is a suspension system in which at least 75% of the spring effort is caused by an air spring. An air spring is a spring operated by air or other compressible fluid under pressure.

Rigid Vehicles – Two axles

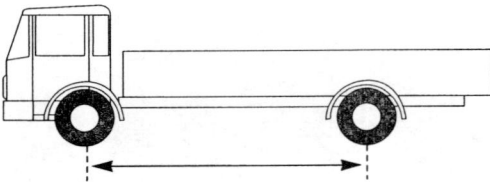

	Max GVW
Less than 2.65m	14.23t
At least 2.65m	16.26t
At least 3.00	17.00t

Rigid Vehicles – Three axles

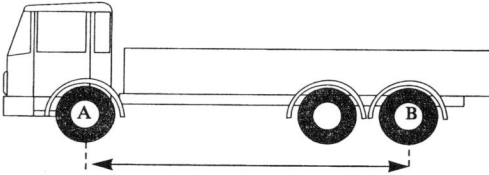

A–B	Max GVW
Less than 3.0m	16.260t
At least 3.0m but less than 3.2m	18.290t
At least 3.2m but less than 3.9m	20.330t
At least 3.9m but less than 4.9m	22.360t
At least 4.9m	25.000t
At least 4.9m but less than 5.2m	25.000t
At least 5.2m	26.000t*

* Road friendly suspension and twin tyres on drive axles.

Rigid Vehicles – Four or more axles

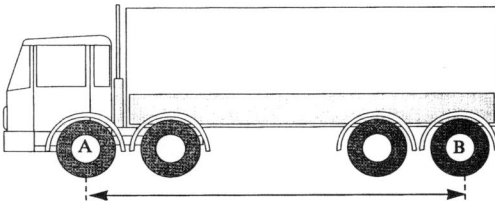

A–B	Max GVW
At least 4.9m but less than 5.6m	25.000t
At least 5.6m but less than 5.9m	26.420t
At least 5.9m but less than 6.3m	28.450t
At least 6.3m	30.000t
At least 5.2m but less than 6.4m	The distance in metres between the foremost and rearmost axles multipled by 5.000 and rounded up to the next 10kg *
At least 6.4m	32.000t *

* Road friendly suspension and twin tyres on drive axles.

Tractor Units for Articulated Vehicles – Two axled unit

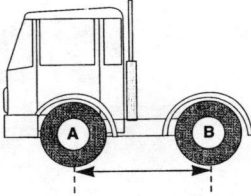

A–B	Max GVW
At least 2.0	14.23t
At least 2.4	16.26t
At least 2.7	17.00t

Tractor Units for Articulated Vehicles – Three or four axled unit

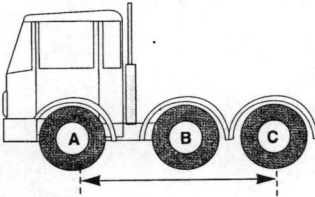

B not more than	A–C	Max GVW
8.39t	At least 3.0	20.33t
8.64t	At least 3.8	22.36t
10.50t	At least 4.0	22.50t
9.15t	At least 4.3	24.39t
10.50t	At least 4.9	24.39t

Articulated Combinations

Two Axle Tractor Unit with Single Axle Trailer

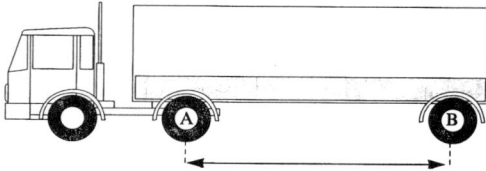

A–B	Max GVW
At least 2.0m	20.33t
At least 2.2m	22.36t
At least 2.6m	23.37t
At least 2.9m	25.00t
At least 3.2m	26.00t*

*Road friendly suspension or international journeys.

Two Axle Tractor Unit with Two Axle Trailer

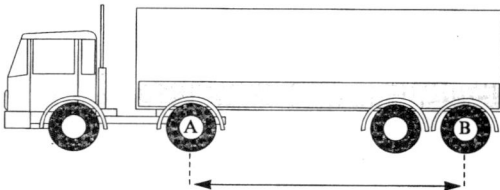

A–B	Max GVW
At least 3.2m	25.41t
At least 3.5m	26.42t
At least 3.8m	27.44t
At least 4.1m	28.45t
At least 4.4m	29.47t
At least 4.7m	30.49t
At least 5.0m	31.50t
At least 5.3m	32.52t
At least 5.5m	33.00t*
At least 5.8m	34.00t*
At least 6.2m	35.00t*

* Road friendly suspension or international journeys.

Two Axle Tractor Unit with Three Axle Trailer

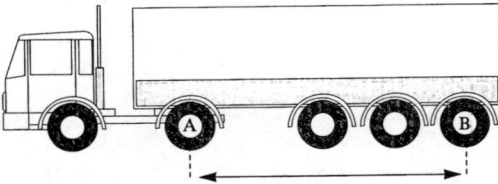

A–B	Max GVW
At least 5.5m	33.00t
At least 5.8m	34.00t
At least 6.2m	35.00t
At least 6.5m	36.00t
At least 6.7m	37.00t
At least 6.9m	38.00t

Three Axle Tractor Unit with One, Two or Three Axle Trailer

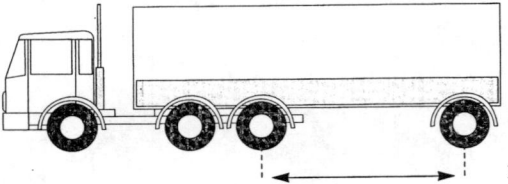

One or Two Axle Trailer	Max GVW
At least 2.0m	20.33t
At least 2.2m	22.36t
At least 2.6m	23.27t
At least 2.9m	24.39t
At least 3.2m	25.41t
At least 3.5m	26.42t
At least 3.8m	27.44t
At least 4.1m	28.45t
At least 4.4m	29.47t
At least 4.7m	30.49t
At least 5.0m	31.50t
At least 5.3m	32.50t

Road-rail vehicles
Certain six-axled articulated and drawbar
vehicle combinations are permitted to
operate at up to 44 tonnes gross weight when
delivering or collecting containers and swap
bodies.

Two or Three Axle Trailer

At least 5.4m	33.00t
At least 5.6m	34.00t
At least 5.8m	35.00t
At least 6.0m	36.00t
At least 6.2m	37.00t
At laest 6.3m	38.00t

MAXIMUM VEHICLE LENGTHS

Rigid vehicles	12m
Articulated vehicles	16.5m
Rigid and trailer	18.75m

Overhang
Rigid goods vehicles – 60 per cent of the
wheelbase.

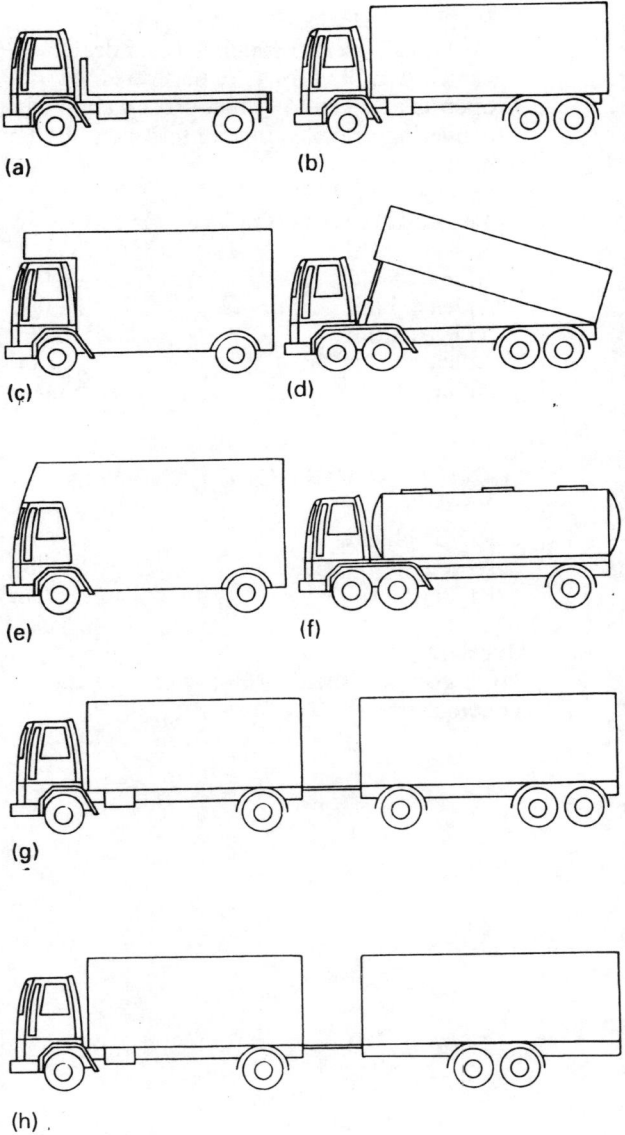

(a)

(b)

(c)

(d)

(e)

(f)

(g)

(h)

Figure 74
Rigid Vehicles

22
Truck Driver of the Year Competition

The Truck Driver of the Year Competition (formally the Lorry Driver of the Year) is an annual event organised by professionals for professionals.

The event is aimed to promote safe, skilful driving in conjunction with driving knowledge and best practices.

The competition is organised at local and national levels.

The local or regional centres as they are known, each have an organising committee drawn from local people who are actively involved in the road transport and distribution industry, ie road safety officers, the local police, professional driving groups, driving associates and many more.

The events have been running for over 40 years and have always attracted a high calibre individual driver and organisation.

The competition is open to all professional drivers and goods vehicle operators who meet the entry requirements as laid down by the Truck Driver of the Year organisers.

The entry requirements are mainly concerned with type or classification of the entry and you are able to obtain this information from the Truck Driver of the

Year Competition, P O Box 881,
Aberystwyth, Dyfed, SY24 5LG.

Competitors may enter only one local contest in any one area. However, a company may enter different competitors at a number of local/regional contests.

Drivers entering the Competition must be in possession of a valid, full, appropriate driving licence covering the class or type of vehicle being used in the Competition.

Drivers will also be required to sign a declaration that during the last 12 months leading up to the competition they:

- have been free from blameworthy accidents
- have had a full valid licence for the last 12 months
- have not had an endorsement on the driving licence
- have not had their driving licence suspended.

Vehicles used in the competition must meet all legal requirements, and comply with all relevant construction and use regulations.

Demountable bodies or container carriers on a vehicle will be deemed to be part of the vehicle.

Vehicles equipped with any form of CB, telephone or other such equipment must have it immobilised or removed whilst competing in any part of the competition.

Competing vehicles are divided into three main categories:

- Rigid vehicles
- Articulated vehicles
- Rigid vehicles and drawbar combination.

Rigid vehicles

Having two axles and:

- Over 3.5 tonnes GVW up to and including 7.5 tonnes GVW (Non LGV)
- Not less than 5.5m, up to and including maximum legal length (12m)

- A valid full driving licence.

Having 2 or more axles and:
- Over 7.5 tonnes GVW
- Not less than 5.5m, up to and including maximum legal lengths (12m)
- Relevant LGV driving licence.

Articulated vehicles
Having 3 or 4 axles and:
- Over 7.5 tonnes GVW
- Not less than 9m, up to and including maximum legal length (16.5m)
- Relevant LGV driving licence.

Having 5 or more axles and:
- Over 7.5 tonnes GVW
- Not less than 9m up to and including maximum legal length (16.5m)
- Relevant LGV driving licence.

Rigid vehicle and drawbar combination
- The trailer having a steerable axle
- Over 7.5 tonnes gross train weight
- Not less than 14m up to and including maximum legal length (18.75m)
- Relevant LGV driving licence.

- The trailer having non-steerable axle(s)
- Over 7.5 tonnes gross train weight
- Not less than 14m up to and including maximum legal length (18.75m)
- Relevant LGV driving licence.

The competition itself is in five parts:
1. Scrutinising
2. Practical road craft
3. Theory – *Highway Code* and road markings
4. Judgement of speed and distance
5. Vehicle manoeuvring skills

Part One – Scrutinising
All competitors will be issued with a vehicle number card which will be required to be

fixed to the front and rear of the vehicle.

Vehicles will then be scrutinised to ascertain that they are entered into the appropriate class and that they comply with all the regulations governing the competition.

Part Two – Practical road craft

Competitors must follow the route indicated by the map or route card which will be supplied by the organisers. Competitors are advised that it will be to their advantage to be accompanied by one adult passenger for this part of the competition. At certain points on the route, not previously disclosed, observers will be posted. Penalties will be imposed for breaches of the *Highway Code* and for poor driving technique.

Part Three – Theory: *Highway Code* and road markings

Questions will be asked on the current edition of the *Highway Code* and standard road signs.

Part Four – Judgment of speed and distance

Competitors will be examined on their ability to judge speed and distance in accordance with the advice given in paragraph 57 of the current edition of the *Highway Code*.

Part Five – Vehicle manoeuvring skills

Competitors will undertake four manoeuvring tests. The four tests will be selected from those illustrated in the national regulations, and may vary from centre to centre. Before undertaking the tests, competitors must obscure all windows to the rear of the cab doors, and any forward or side facing windows below the base level of the main windscreen to the satisfaction of the organisers. No passengers may be carried during manoeuvring tests.

A competitor is permitted to open and look out of the driver's side door only when reversing. Apart from this, all doors must be closed during all tests, but windows may be open. Throughout the tests the competitor must remain in a normal driving position and be able to operate all vehicle controls.

The following illustration has been selected from previous manoeuvring tests as used in the Truck Driver of the Year Competition.

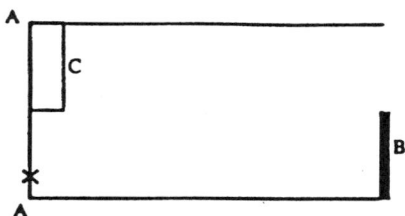

TEST

Start with the offside front wheel at point X on line AA. Drive forward and stop with the foremost part of the vehicle as close as possible to, but not touching, barrier B. When instructed, reverse and stop with the rearmost, outermost wheels in box C.
One forward move and one reverse move are permitted, both of which must be continuous.

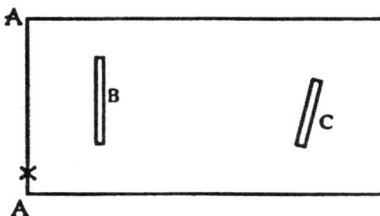

TEST

Start with the offside front wheel at point X on line AA. Drive forward and stop with all rearmost wheels on target B.
When instructed, drive forward and stop with the foremost part of the vehicle directly over target C.
Two forward moves are permitted, both of which must be continuous.

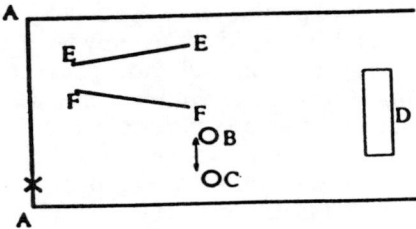

TEST

Start with the front offside wheel at point X on line AA.
Signal to the marshal to move cone B away from cone
C to leave the minimum gap through which the vehicle
will pass without touching.

When instructed, drive forward between cones B and
C, and stop with the whole of the width of the front of
the vehicle over target D.

When instructed, reverse to stop with the rearmost,
outermost, nearside wheel as close as possible to, but
not touching, kerb EE, and with the nearmost,
outermost, offside wheel as close as possible to, but not
touching, kerb FF. One forward and one reverse move
are permitted, both of which must be continuous.

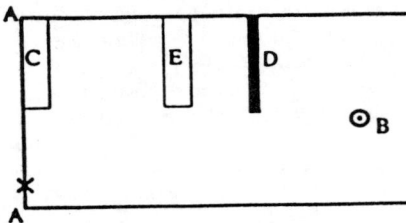

TEST

Start with the front offside wheel at point X on line AA.

When instructed, drive forward and stop with the
centre of the front of the vehicle as close as possible to,
but not touching, pole B.

When instructed, reverse and stop with both rearmost,
outermost wheels in box C.

When instructed, drive forward and stop with the front
of the vehicle as close as possible to, but not touching,
barrier D.

When instructed, reverse and stop with both front
wheels on target E.

Two forward and two reverse moves are permitted,
each of which must be continuous.

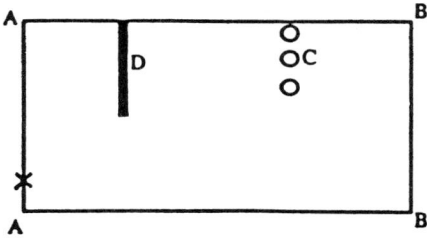

TEST

Start with the offside front wheel at point X on line AA.
Drive forward and stop before any wheel touches line
BB.

Reverse to park between cones C and barrier D with all
nearside wheels as close as possible to, but not
touching, the nearside kerb.

Two forward moves and one reverse move are
permitted, each of which must be continuous.

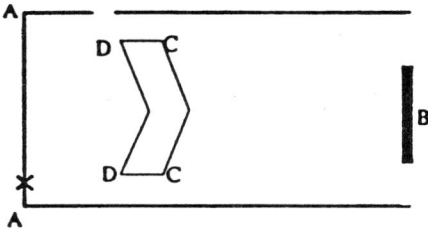

TEST

Start with the offside front wheel at point X on line AA.
Drive forward and stop with the foremost part of the
vehicle as close as possible to, but not touching, barrier
B.

When instructed, reverse and stop with the rearmost
outermost wheels between lines CC and DD.

One forward move and one reverse move are
permitted, both of which must be continuous.

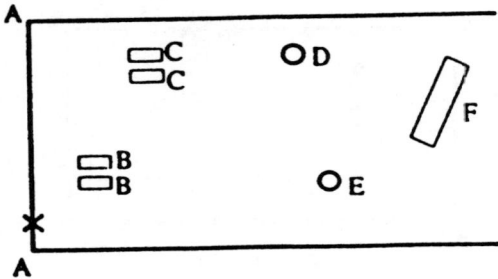

TEST

Start with the front offside wheel at point X on line AA.
When instructed, drive forward to pass with the front
offside wheel only between blocks BB, to pass with
front nearside wheel only between blocks CC, to pass
with front and rearmost wheels only centrally between
cones D and E, and stop with full width of the rear
vehicle over target F.

One forward move is permitted, which must be
continuous.

Figure 75

23
Waiting, Parking and Loading

Parking
Parking Restrictions
Parking Without Lights
Overnight Parking
Red Routes In London
Breakdown In Restricted Areas
Road And Kerb Signs
Dangerous Loads
Kerbside Restrictions

PARKING

When parking a vehicle the driver must always consider two things:
1. Is it *safe*?
2. Is it *legal*?

In certain circumstances it may be legal to park but due to other circumstances it may not be safe, in which case a driver would leave himself wide open to prosecution.

For example:
The illustration shown in Figure 76 shows a truck parked safely and legally.

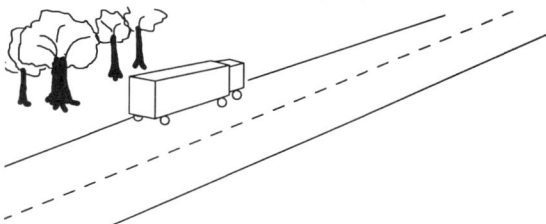

Figure 76

By adding some roadworks to the picture as shown in Figure 77 the parking of the truck has become unsafe and the driver is now wide open for prosecution.

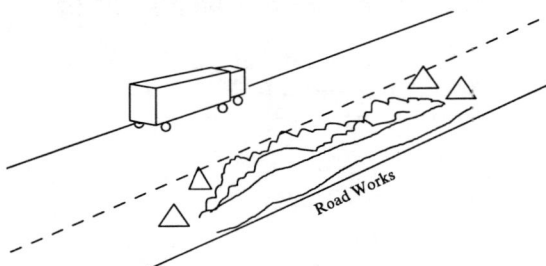

Figure 77

Drivers of large goods vehicles should, wherever possible, use authorised parking facilities such as official car/lorry parks, motorway service areas, lay-bys, truck stops and so on.

Drivers loading or unloading goods are not always subject to the same restrictions as drivers who are parking their vehicles.

PARKING RESTRICTIONS

It is an offence to leave a vehicle unattended with the engine running except for the purpose of using ancillary equipment and even then the parking brake must be on and the vehicle must not be left on a road in such a way as to cause an obstruction or inconvenience to others.

Trailers and semi-trailers must not be left on the road detached from the prime mover unless the trailer brake is on or the wheels are locked and are prevented from movement.

PARKING WITHOUT LIGHTS

Small goods vehicles, not exceeding 1525 kgs unladen weight, may park on the road at night without lights, including roads that have been marked as designated parking areas providing:

- the road is subject to a 30 mph speed limit or less
- no part of the vehicle is within 10 metres of a road junction and there are no waiting restrictions in force
- the vehicle is parked close to the kerb on the nearside, except when the vehicle is parked in a recognised parking place or in a one way street.

Large goods vehicles exceeding 1525 kgs unladen weight and any vehicle to which a trailer is attached, including articulated vehicles, *must* have sidelights on (front side and rear) when parked on any road during the hours of darkness.

Note: Drivers must be mindful not to show white lights facing oncoming traffic, vehicles must be parked with the nearside close to and parallel to the roadside kerb, so that the red rear light can be clearly seen by other vehicles travelling in the same direction.

OVERNIGHT PARKING

Overnight parking for large goods vehicles has always presented the long distance truck driver with the problem of where to park his/her vehicle safely for the night.

Both driver and haulier are demanding better facilities. The driver who has the final responsibility for the safe parking of the vehicle and its load will normally make arrangements in advance. Many large industrial parks and industrial centres are offering safe parking for vehicles and in some cases are also providing the driver with the facilities that he/she requires. In addition to industrial park and distribution centres, some large transport organisations are also able to offer secure parking.

Many local authorities are controlling the parking of goods vehicles overnight by providing designated parking streets.

Drivers must look for and abide by the local parking signs. Failure to do so will lead to action being taken.

To ensure you do not fall foul of the local restrictions, enquire at the local Police Station which will hold a list of local overnight parking facilities suitable for large goods vehicles.

RED ROUTES IN LONDON

The Red Route Scheme started in January 1991 covering many roads in London. The rule is quite simple 'NO STOPPING' for all or part of the day to improve traffic flow and cut accidents.

The routes are clearly marked and there are double, single and hatched red lines to indicate the different zones:

Double red line indicates:
no stopping at all at any time
Single red line means:
no stopping at certain times (mainly during the working day Monday–Friday)
Hatched red lines (or boxes) indicate:
parking bays with limited time

For drivers of goods vehicles, signs will indicate where loading/unloading is permitted during certain hours otherwise you are not permitted to stop. There are the usual exceptions for buses and licensed taxis,

disabled persons and in emergencies, but you
need to be aware of the places and times of
unloading, and have clear instructions of
what to do if you arrive outside the
permitted hours.

Enforcement authorities have stated
publicly that they intend to police the new
Red Routes more effectively than they have
been able to patrol the normal waiting
restrictions.

BREAKDOWN IN RESTRICTED AREAS

Should your vehicle break down where a
waiting restriction is in force, and it is
necessary for you to leave it unattended to
summon assistance, you should notify the
police or a traffic warden. The police are
empowered to remove any vehicle which is
causing an obstruction, or is left in a
dangerous position, or is contravening
waiting, loading/unloading restrictions. It is
therefore advisable to leave a note on the
windscreen indicating the problem and the
action that is being taken.

ROAD AND KERB SIGNS

Yellow lines painted along the gutter area of
the road or carriageway indicate no waiting.

Yellow strips painted on the kerb indicate
no loading or unloading.

In both cases a nearby plate will give
further details.

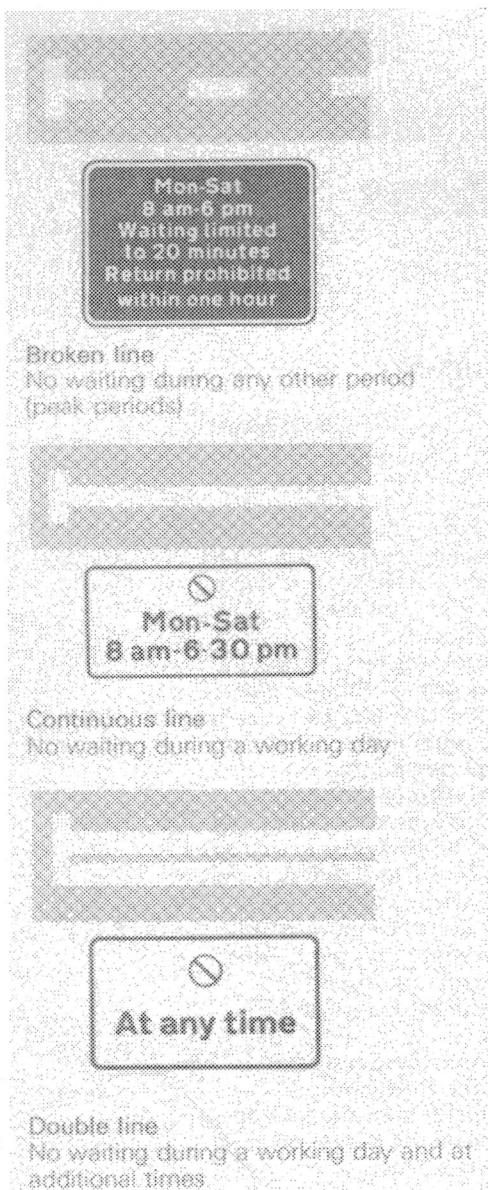

Mon-Sat
8 am-6 pm
Waiting limited
to 20 minutes
Return prohibited
within one hour

Broken line
No waiting during any other period
(peak periods)

Mon-Sat
8 am-6·30 pm

Continuous line
No waiting during a working day

At any time

Double line
No waiting during a working day and at
additional times

Figure 78

Zone exit sign

Figure 79

Zone entry sign
Where the restriction applies throughout a zone, zone entry signs mark the limits of the restricted area

Figure 80

Advisory lorry routes
Where a port or industrial centre is not directly on a primary route, the white lorry symbol may be used to mark the most suitable route for heavy goods vehicles to and from the primary route

There are three types of traffic sign relating to these regulations:

- *Controlled zone boundary signs:* these signs indicate the boundary and hours of operation of the controlled zone. They will appear at every entry to, and exit from, the controlled zone (see Figure 81)
- *Authorised parking place sign:* these signs define the limits of the overnight parking bays in the designated streets (see Figure 82).

^
Figure 81
A. Sign indicating the entrance to a Controlled Parking Zone applying to buses and heavy vehicles and trailers over 3.5T GVW

B. Sign indicating the exit from a Controlled Parking Zone applying to buses and heavy vehicles and trailers over 3.5T GWV

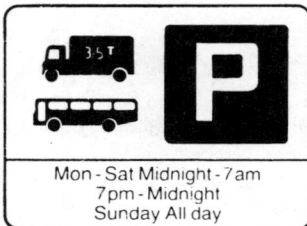

Figure 82
Sign indicating an Authorised Parking Place reserved for buses and heavy vehicles and trailers over 3.5T GVW

KERBSIDE RESTRICTIONS

Yellow stripes painted on the kerb indicate
loading and unloading restrictions. A nearby
plate will give details.

One stripe
No loading or unloading during peak periods
and at other times

Two stripes
No loading or unloading during a
working day

Three stripes
No loading or unloading at any time (*this sign
to be abolished from 1999*)

Figure 83

- You should obtain permission from the police if you need to deliver or collect where such restrictions are in force
- Where a waiting restriction is in force you may stop to load or unload for 20 minutes only. If you need a longer time you should obtain permission from the police. If you are not loading or unloading you must observe the waiting restrictions
- You can load or unload in a vacant parking meter bay for 20 minutes without charge, but a police officer or traffic warden is entitled to move you on.

24
Road Signs

Warning Signals Near Railways

A wide range of traffic signs which have been gradually introduced on our roads are illustrated here. All will be relevant to the driving of LGVs.

STOP sign
Stop and give way.

Mini roundabout
Give way to traffic from the immediate right.

Loose chippings

Staggered junction
This and other versions of junction signs indicate the priority through the junction by the thickened line.

T-junction

Crossroad

Side road

Tunnel
Indicates the road ahead entering a tunnel – look out for head room.

Steep hill down
The gradient should be included on this sign. The sign is often accompanied by a plate: 'low gear now'. Gradients may also be shown as a ratio, eg 20% = 1:5

Steep hill up

On relatively steep hills on motorways an extra lane is sometimes provided so that slow-moving vehicles need not impede faster traffic.

The **bus lane sign** and **road marking**
In some towns specific traffic lanes may be reserved for buses. Goods vehicles may cross the white line and enter the bus lane to stop or unload goods, but only at times when there is no restriction on loading.

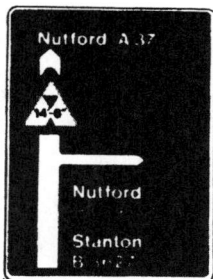

An **advance direction sign** may indicate a restriction and offer an alternative route.

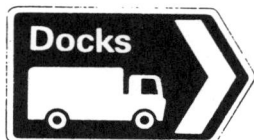

Advisory lorry routes
Where a port or an industrial centre is not directly on a primary route the white lorry symbol may be used to mark the most suitable route for LGVs to and from the primary route.

No vehicle or combination of vehicles over length shown

No vehicles (including load) over weight shown (in tonnes)

No goods vehicles over maximum gross weight shown (in tonnes)

Axle weight limit in tonnes

No vehicles over width shown

No vehicles with over 12 seats except regular scheduled school and works buses

No vehicle over height shown

End of goods vehicle restriction

WARNING SIGNALS NEAR RAILWAYS

At automatic half-barrier level crossings drivers of large or slow vehicles *must* telephone both before and after crossing.

'Large' means over 55' long or 9'6" wide or 38 tonnes total weight. 'Slow' means 5 mph or less.

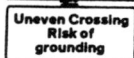

Where there is danger of grounding you may see this warning sign at the last road junction before the crossings and on the approaches to these and other types of crossing. Drivers of vehicles with low ground clearance should not try to cross until they have made sure that there is no risk of grounding on the crossing. *If in doubt drivers should telephone the signalman* – or speak to him direct if he is at the crossing.

Overhead electric cables
The plate indicates the maximum safe height for vehicles.

'Count-down' markers approaching a concealed level crossing. Each bar represents one-third of the distance from the warning sign to the crossing.

25
The *Highway Code*

Questions And Answers

Most people read the *Highway Code* just before a driving test and then put it away and forget it. Many drivers are convinced that they fully understand the *Highway Code*. In reality, however, very few are entitled to such confidence.

This chapter offers you the opportunity to test *your* knowledge. Some general questions and answers about driving are included.

QUESTIONS AND ANSWERS

What is the maximum speed limit for a large goods vehicle on a motorway?

60 miles per hour.

What is the colour coding of cats' eyes on a motorway?

Red, white, amber. Red changes to green at intersections.

Large goods vehicles must not use the third lane of a three-lane motorway. Give two exceptions to this rule.

When passing a wide load.
When passing road works.

*What are the eight rules that should be
observed when driving in fog?*

- Slow down, and keep a safe distance
  from the vehicle in front
- Do not drive on the tail lights of the
  vehicle in front
- Keep a constant check on your speed.
  Don't increase it to get away from a
  vehicle that is too close behind you
- *Remember:* it may take longer for a *large*
  goods vehicle to stop
- Observe *all* warning signals
- Headlamps, foglamps, and rear lamps
  must be used – See and be seen
- Keep all windows, lights, reflectors and
  windscreen clean
- Allow extra time for the journey.

*Cross winds on a motorway are a serious
danger for high-sided vehicles – what should
you do?*

Slow down, and consider leaving the
motorway.

*What are the three signs displayed on the
motorway roadside signals (motorway central
reserve)?*

- Temporary maximum speed
- Lane closed ahead
- End of restriction.

*What are the five signs displayed on the
overhead gantries (motorway overhead
signals)?*

- Temporary maximum speed
- Change lane
- Leave motorway at next exit
- Stop
- End of restriction.

At some automatic level crossings there are large rectangular signs that say, 'Drivers of large or slow vehicles must phone and get permission to cross'. What is defined as large or slow?

- Large – over 16–18 m (55 ft) long
  over 2–9 m (9 ft 6 in) wide
  over 38 tonnes gross weight
- Slow – 5 mph or less.

What is meant by grounding?

The bottom of the trailer catches the peak of an uneven road surface. This may happen with certain types of trailer, particularly low loaders.

If there is a danger of grounding what sign may you see?

- Uneven road surface
- Hump bridge.

What are the general rules covering the use of all railway crossings?

- Never drive nose to tail over any level crossing
- Never drive on to a crossing unless it is clear on the other side
- Never stop on or immediately beyond a level crossing.

What are the rules concerning a pedestrian crossing marked with zig-zag lines?

- Do not wait or park on the zig-zags
- Do not overtake a moving vehicle that is nearest to the crossing on the approach side
- Give way to pedestrians on the crossing.

What is the sequence of light signals at a pedestrian crossing?

Green – Amber on its own – Red – Flashing amber – Green.

What does flashing amber mean?

Give way to pedestrians who are already crossing, but if your way is clear you may proceed.

What rule should you observe at a yellow box junction?

Do not enter unless your exit is clear, except when you are turning right and are only prevented from doing so by oncoming traffic.

What are the rules concerning double white lines in the centre of the road?

Where the line nearest to you is broken you may cross provided it is safe to do so, but where the line nearest to you is continuous you must not cross except under special circumstances.

Can you park on a road where there are double white lines down the centre?

No.

What are the signs or markings which prohibit waiting and parking?

- No waiting sign
- Clearway sign
- Yellow lines in gutter
- Double white lines in road.

Give three occasions when the horn should not be used.

- Between 11.30 pm and 7.00 am in a built-up area
- When the vehicle is stationary (except in an emergency)
- When used as a rebuke to other road users.

You should know the answers to these questions, and you should be able to recognise *immediately* the following road signs:

No goods vehicles or combination of vehicles over length shown.

No vehicles (including load) over weight shown.

No goods vehicles over gross weight shown (including load).

No vehicles over axle weight shown.

No vehicles over width shown.

No vehicles with 12 or more seats, except
regular scheduled school and works services.

No vehicles over height shown.

26
The Channel
Tunnel

INTRODUCTION

With the opening of the Channel Tunnel in 1994, many drivers who normally drive short, middle or long distance in the UK may find themselves using the Le Shuttle freight service through the Tunnel and extending these journeys by driving in and around Europe as part of their normal working day.

European driving is still something of a specialism undertaken by hauliers who specialise in European transport and distribution. As time goes on, however, more and more local hauliers may well expand their scheduled delivery routes, taking into account the European trade market.

THE TERMINALS

The Euro terminals are located at Cheriton on the outskirts of Folkestone on the UK side of the English Channel and at Coquelles on the outskirts of Calais on the French side.

Each terminal covers hundreds of acres offering almost every facility that a truck driver could require. Improved motorway network links and high-speed rail links on both sides connect France and the UK to a very comprehensive transport system.

THE TUNNEL

The tunnel itself is actually made up of three different tunnels: Two 'running tunnels' and a central service tunnel. One of the largest European engineering projects of its kind, the tunnel is about 50 kilometres long and runs between 25 and 45 metres beneath the sea bed.

The two running tunnels are 7.6 metres in diameter, each containing a single railway track with interconnecting tunnels to link them. A sophisticated air-duct system reduces the build-up of air pressure in front of the high-speed trains.

All three tunnels have immensely strong linings made from cast iron or reinforced concrete segments; the concrete segments weigh up to 10 tonnes each with cement grout pumped behind the lining to provide a final seal between the tunnel wall and the tubular tunnel lining. (The concrete is up to nine times stronger than the concrete used to build a nuclear reactor.)

A comprehensive drainage, ventilation and signalling system is controlled and monitored from each terminal, in addition to localised maintenance via the central service tunnel.

THE TRAIN AND TRACK

Le Shuttle is like a road train on railway lines stretching up to half a mile and it weighs

*somewhere in the region of 3000 tonnes
GVW. It comprises seven individual rolling
stock units with a prime mover (the
locomotive) at the front, and a second
locomotive at the rear, with a drivers' club
car, four vehicle loading/unloading ramps
and two LGV vehicle carriers between.*

THE PRIME MOVERS

The prime mover is an all-electric train
powered by overhead pantograph cables.
The power, which is computer controlled, is
purchased from either UK or French grids
according to which is cheapest at the time,
but Eurotunnel also has its own power
station for use as required.

THE OPERATION

The operating principles involved in moving
an LGV from the UK to France or vice versa
are simple: a driver will arrive at the
respective terminal and pass through a series
of high-speed checkpoints and facilities and
drive the vehicle on to a train.

The train will take its cargo of vehicles
and passengers through the tunnel to the
opposite terminal where the driver will
simply drive the vehicle off the train directly
on to the respective motorway or road links.
There is no need or requirement for a driver
to pass through any additional checkpoints
or controls.

The text below describes the driver's
journey in more detail, identifying all
facilities, checkpoints and controls
encountered as he/she travels from the UK
to France.

The approach road to the UK terminal is
the M20. Road signs and markings clearly

identify the driver's route. The first facility available to the driver is the truck stop, or to use its correct title The Channel Stop Le Shuttle, located just outside the terminal off the M20. This facility is exclusive to truck drivers, and is open 24 hours a day all year round. A large complex offering first-class secure parking for up to 273 road trains, it has its own vehicle service station for LGVs only. The service area is similar to the motorway service area offering oil, derv, water, air and all the normal forecourt refinements a driver would expect – including a well-stocked shop.

The main facility is the 'Drivers Motec', which is a large single-storey building where drivers can relax in comfort. It is similar to a motorway service station, though with an international flavour. *The reception* is open 24 hours and reception staff are multi-lingual, friendly and helpful. They are able to organise external facilities such as a doctor, dentist, hospital, taxis and so on. They are also able to arrange external facilities for the vehicles such as windscreen replacement, tyre service and vehicle repair services. The facility offers a first-class *restaurant* serving excellent food at very competitive prices.

Drivers' rest rooms are quiet and comfortable, and are also well equipped with *TV rooms, games rooms* and a *bar* serving alcoholic and non-alcoholic drinks, again at very competitive prices.

Accommodation is plain, clean and comfortable with separate shower and toilet facilities. Reasonably priced, it includes overnight parking. Although pre-booking is not required, drivers may pre-book their accommodation if they wish (tel: 01233 501122). The truck stop is only a short journey from the terminals and, as drivers arrive at the terminal, they will follow their own road signs as they enter the segregation lanes for cars, buses and trucks.

The segregation lanes lead drivers to their respective *toll booths*. The vehicle then passes between and under height- and width-monitoring poles equipped with 'magic eye' indictors which will sound if the vehicle is too high or too wide to be allowed on the train.

The maximum limits on height, length, width and weight are:

Height	4.2 metres
Length	18.5 metres
Width	2.5 metres
Weight	44 tonnes GVW.

Once through the height and width test, the driver would show a pre-paid *rapid service pass* that generates an invoice for the crossing. A rapid service pass is negotiated by the haulier via the Le Shuttle UK freight sales team in advance. The cost of these passes varies according to the frequency or number of crossings the haulier is likely to make over a given period. (For more details about the rapid service pass the haulier should contact freight sales on 01303 273300.)

As drivers clear the toll booths they are directed towards UK passport control, passing the *freight drivers' amenity building* where drivers may wish to take advantage of the services offered.

The amenity building is exclusive to truck drivers and offers the chance to buy duty free or gifts. It is a well-equipped building with telephone and fax facilities, rest rooms, wash rooms, currency exchange and an information desk.

Drivers may also purchase light refreshments, post cards and stamps before moving on to passport control. *Passport control* is similar to the toll booth: drivers do not have to leave the vehicle in order to have their passport checked. French passport control, where the passport is checked again, is just a few yards away.

During the short journey between the two passport control points, vehicles are selected at random and diverted to a building called '*Euro-Scan*'. At this intersection the vehicle is driven on to a conveyor belt platform while the driver walks the covered pathway to the other end of the building. The vehicle will enter the building moving along the conveyor belt (not driven) in order for the vehicle and its cargo to be x-rayed.

This security check is to identify any potential hazards that may cause damage to the tunnel or any contraband.

The whole process (UK passport control, Euro-Scan and French passport control) takes only minutes. The final control before the vehicle is driven on to the train is the *platform allocation booth* where drivers are allocated a specific lane to follow which takes them to the correct position on the platform. The vehicle can then be driven on to the train's loading ramp and into the vehicle carrier wagons.

Loading ramps are flat-bed platforms on wheels positioned behind and in front of the two vehicle carriers. Platform staff are in attendance to guide drivers on to the loading ramps and safely into the vehicle carriers.

There are two vehicle carriers to each train and each carrier is capable of receiving up to 14 vehicle road trains. As the vehicle is secured in the vehicle carrier the driver will leave his/her vehicle.

Because of the length of the train (half a mile), a platform mini-bus service will pick up the drivers and take them to the drivers' club car. An exclusive facility for LGV drivers only, the club car is a cross between a luxury railway carriage and a passenger aircraft.

As part of the Le Shuttle freight service, drivers are served at no cost with newspapers, magazines and hot and cold meals with a choice of hot or cold drinks. This enables drivers to take an official break as the train speeds along the 35-minute journey to the opposite platform on the French side.The total journey time from the toll booth in the UK to the final drive in France is 80 minutes.

When in full swing theLe Shuttle freight service operates every 15 minutes, 24 hours a day, 7 days a week.

INTERNATIONAL ROAD SIGNS

These signs are in general use in Europe but may not be familiar to British drivers.

GENERAL

Priority Road

End of priority

Intersection
(priority rule applies)

Intersection
with tramway

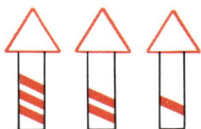
Approaches to level crossings

Use of audible
warning devices
prohibited

AUSTRIA

Diversion

Tram turns
at yellow or red

Federal road
with priority

Federal road
without priority

No entry for
lorries with trailers

Prohibited to
vehicles carrying
dangerous goods

Buses only

U-turn compulsory

BELGIUM

Turn right or left

No parking from
1st to 15th of month

No parking from
16th to end of month

DENMARK

Sight-seeing

Maximum weight
of vehicle or trailer

Maximum weight
of vehicle and trailer

Maximum weight
on double axle

Compulsory
slow lane

Recommended
speed in a bend

Traffic merges

FINLAND

Diversion due
to road works

Prohibition applied
between 0800 and
1800 hrs Mon-Sat

Prohibition applies
between 0800 and
1400 hrs (Saturday)

Prohibition applies
between 1000 and
1400 hrs Sunday

FRANCE

Keep well over
to the right

Bus lane

Easily
inflammable forest

No parking from
1st to 15th cf month

No parking from
16th to end of month

Fortnightly parking
or alternate sides

Relief route

Holiday route

WEST GERMANY

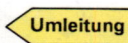
Diversion

Street lights
not on all night

Tram or bus stop

Emergency diversion
for motorway traffic

Recommended
speed range

HUNGARY

Diversion

Route for heavy vehicles

Lane reserved for buses from 0700 to 1900 hrs

ITALY

Use chains or snow tyres from Km 174

Traffic in parallel lanes

Lane reserved for slow vehicles

Track for motorcycles

Road for motor vehicles

Restricted parking

Overtaking by vehicles with trailers prohibited

No entry for pedestrians

Stop when meeting public transport bus on mountain road

Easily inflammable forest

NETHERLANDS

Cycle track (mopeds)

End of built-up area

End of B road

Crossing for cyclists and moped riders

Compulsory route for vehicles with dangerous goods

road (width and axle weight limits)

No vehicle or combination over length shown

Please watch, be extra vigilant for cyclists.

PORTUGAL/SPAIN

End of parking
prohibition

Recommended
maximum speed

Take care (yellow
or white triangle)

Turning permitted

Tourist office

Sight-seeing

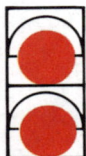
No entry

Compulsory lane
for motorcycles

Compulsory lane
for lorries

Easily
inflammable forest

SWEDEN

Maximum weight
on double axle

Passing place
(on narrow roads)

Tunnel

Slow lane

SWITZERLAND

Tunnel
(lights compulsory)

Heavy coaches
prohibited

Motorway

Semi-motorway

Flashing red light
(level crossing)

Alternately
flashing lights
(level crossing)

Parking disc
compulsory

Passing place
for lorries

Slow lane

Trailers prohibited

Animals prohibited

Postal vehicles
have priority

27
Motorway Services

M25, M1, M2, M3, M4, M5, M6, M61
M62, M27, A1(M), M40, M8, M9, M11,
M42 M74, M74(M) AND M90.

The author wishes to formally acknowledge
The Automobile Association for their kind
permission in providing the Motorway
Service Area Listing.

M25

London Orbital

Clacket Lane
Junctions 5–6
Tel: 01959 565575
Roadchef
Restaurant. Fast food – Burger King.
Truckers restaurant. Autobank. 'Le Shuttle'
Centre. Shop. Shower facilities.

South Mimms
Off junction 23
Tel: 01707 646333
Welcome Break
Restaurant. Cafeteria. Fast food. Autobank.
Shop. Accommodation Tel: 0800 850950.

Thurrock
Junctions 30–31
Tel: 01708 865487
Granada
Restaurant. Fast food – Burger King. Shop.
Breakdown and repair facilities.
Accommodation, Tel: 0800 555 300.

M1

London – Leeds

Scratchwood
Junctions 2–4
Tel: 0181 906 0611
Welcome Break
Granary–Restaurant. Cafeteria.
Accommodation, Tel: 0181 906 0611. 24hr
fuel. Shop. Breakdown and repair facilities.

Luton (N)
Junctions 11–12
Tel: 01525 875577
Granada
Transport café. Restaurant. Shop. Fast food
– Burger King (Ltd). Breakdown and repair
facilities. Accommodation,
Tel: 0800 555 300.

Newport Pagnell
Junctions 14–15
Tel: 01908 217722
Welcome Break
Cafeteria. Restaurant. Granary. Breakdown
and repair facilities. Accommodation,
Tel: 01908 610878/0800 850950

Rothersthorpe
Junction 15A
Tel: 01327 78811
Blue Boar
Restaurant. Cafeteria. Shop. Breakdown and
repair facilities.

Watford Gap
Junctions 16–17
Tel: 01327 79001
Blue Boar
Restaurant. Transport café. Shop.
Breakdown and repair facilities.

Leicester Forest East
Junctions 21–22
Tel: (0116) 238 6801
Welcome Break
Cafeteria. Waitress service restaurant. Fast
food. Shop. Breakdown and repair facilities.

Nottingham
Junctions 25–26
Tel: (0115) 932 0291
Granada
Restaurant. Fast food – Burger King. Shop.
Breakdown and repair facilities.
Accomodation, Tel: 0800 555300.

Woodall
Junctions 30–31
Tel: (0114) 248 6434
Welcome Break
Granary. Shop. Autobank. Breakdown and
repair facilities.

Wakefield
Junctions 38–39
Tel: 01924 830371
Granada
Restaurant. Shop. Breakdown and repair
facilities. Accommodation (northbound
only); Tel: 0800 555300.

M2

Rochester – Faversham

Medway Services
Junctions 4–5
Tel: 01634 233343
Granada
Restaurant. Truck driver restaurant, Fast
food – Burger King. Shop. Accommodation,
Tel: 01634 233343.

M3

London – Southampton

Fleet
Junctions 4A–5
Tel: 01252 621656
Welcome Break
Restaurant. Granary. Coffee shop (Ltd).
Cafeteria. Fast food. Shop. Breakdown and
repair facilities. Accommodation, Tel: 0800
850950.

M4

London – Pont Abraham

Heston
Junctions 2–3
Tel: 0181 574 7271
Granada
Restaurant. Fast food – Burger King. Shop.
Breakdown and repair facilities. LPG fuel
available. Accommodation, Tel: 0800 555
300.

Reading
Junctions 11–12
Granada
Restuarant. Fast food – Burger King Shop.

Newbury
Off junction 13
Tel: 01635 248024
Granada
Restaurant. Fast food – Burger King (Ltd).
Shop. Breakdown and repair facilities.

Membury
Junctions 14–15
Tel: 0488 71881
Welcome Break
Restaurant (Ltd). Cafeteria. Fast food (Ltd).
Shop (24 hours Fri-Sun). Breakdown and
repair facilities.

Chippenham
Junctions 17–18
Tel: 01666 837691
Granada
Restaurant. Fast food – Burger King. Shop.
Breakdown and repair facilities.
Accommodation, Tel: 0800 555300.

Magor
Off junction 23
Tel: 01633 881887
First Motorway Services Ltd
Restaurant. Fast food – Burger King. Shop.
Accommodation, Tel: 0800 850950.

Cardiff West
Off junction 33
Tel: 01222 891141
Pavilion
Restaurant. Fast food – Burger King.
Transport café. Shop. Breakdown and repair
facilities. Accommodation, Tel: 0800 555300.

Sarn Park
Off junction 36
Tel: 01656 655332
Welcome Break
Little Chef restaurant (Ltd). Granary
Restaurant. Breakdown and repair facilities.
Accommodation, Tel: 0800 850950.

Swansea
Off junction 47
Tel: 01792 896222
Granada
Restaurant. Fast food – Burger King.
Transport café. Shop. Breakdown and repair
facilities. Autobank. Accommodation, Tel:
0800 555300.

Pont Abraham
Off junction 49
Tel: 01792 884663

Roadchef
Restaurant. Shop. Autobank.

M5

Birmingham – Exeter

Birmingham
Junctions 3–4
Tel: 0121 550 3131
Granada
Restaurant. Fast food – Burger King. Shop.
Breakdown facilities. Accommodation,
Tel: 0800 555300.

Strensham
Junctions 7–8
Tel: 01684 293004
Take A Break
Northbound
Cafeteria. Fast food. Shop. Breakdown and
repair facilities.

Southbound
Cafeteria. Fast food. Shop. Breakdown and
repair facilities.

Michaelwood
Junctions 13–14
Tel: 01454 260631
Welcome Break
Cafeteria. Fast food. Shop. Accomodation,
Tel: 0800 850950.

Gordano
Off junction 19
Tel: 01275 373624
Welcome Break
Restaurant (Ltd). Cafeteria. Fast food.
Shop. Breakdown and repair facilities.
Accommodation, Tel: 0800 850950.

Sedgemoor
Junctions 21–22
Southbound

Tel: 01934 750888
Roadchef
Restaurant. Shop.

Northbound
Tel: 01934 750730
Welcome Break
Restaurant (Ltd). Fast food. Shop.
Accommodation, Tel: 0800 850950.

Taunton Deane
Junctions 25–26
Tel: 01823 271111
Roadchef
Restaurant. Cafeteria. Shop. Breakdown and
repair facilities. Accommodation,
Tel: 01823 332228.

Exeter
Off junction 30
Tel: 01392 436266
Granada
Restaurant, waitress service in hotel.
Restaurant, self-service. Fast food – Burger
King. Shop. Breakdown facilities.
Accommodation,
Tel: 0800 555300.

M6

Rugby – Carlisle

Corley
Junctions 3–4
Tel: 01676 540111
Welcome Break
Restaurant (Ltd). Cafeteria (Ltd). Fast food.
Shop. Breakdown and repair facilities.

Birmingham
Junctions 10A–11
Tel: 01922 412237.
Granada
Restaurant. Fast food – Burger King.
Transport café. Showers. Shop. Breakdown

and repair facilities. Autobank.
Accommodation, Tel: 0800 555300.

Keele
Junctions 15–16
Tel: 01782 626221
Welcome Break
Restaurant (Ltd). Cafeteria. Fast food.
Games area. Shop. Breakdown and repair
facilities.

Sandbach
Junctions 16–17
Tel: 01270 767134
Roadchef
Restaurant. Shop. Breakdown and repair
facilities.

Cheshire (N)
Junctions 18–19
Tel: 01565 634167
Granada
Restaurant. Transport café (Ltd). Fast food
– Burger King. Shop. Breakdown and repair
facilities (northbound only).

Charnock Richard
Junctions 27–28
Tel: 01257 791494
Welcome Break
Restaurant (Ltd). Granary. Shops.
Breakdown and repair facilities.
Accommodation (northbound only),
Tel: 0800 850950.

Lancaster
Junctions 32–33
Tel: 01524 791775
Granada
Restaurant. Truck drivers restaurant (Ltd).
Fast food – Burger King. Shop. Breakdown
and repair facilities. Accommodation
(northbound only), Tel: 01524 791775.

Burton in Kendal
Junctions 35–36
Northbound only
Tel: 01524 781234
Granada
Restaurant. Fast food – Burger King. Shop.

Killington Lake
Junctions 36–37
Southbound only
Tel: 01539 620739
Roadchef
Restaurant. Shop. Breakdown and repair
facilities. Accommodation,
Tel: 01539 621666.

Tebay
Junctions 38–39
Tel: 01539 624511
Westmorland Motorways
Restaurant. Shop. Accommodation,
Tel: 01539 624351.

Carlisle
Junctions 41–42
Tel: 01697 473476
Granada
Restaurant. Fast food – Burger King. Shop.
Breakdown and repair facilities.
Accommodation,
Tel: 0800 555300.

M61

Manchester – Preston

Rivington
Junctions 6–8
Tel: 01204 468641
First Motorway Services Ltd
Restaurant. Transport café (Ltd). Fast food
(Ltd). Shop. Breakdown and repair facilities.

M62

Liverpool – Humberside

Burtonwood
Junctions 7–9
Tel: 01925 651656
Welcome Break
Restaurant (Ltd). Fast food. Shop.
Breakdown facilities. Accommodation
(westbound only), Tel: 01925 651656.

Mancheter (N)
Junctions 18–19
Tel: 0161 643 0911
Granada
Restaurant. Fast food (Ltd). Shop.
Breakdown and repair facilities.
Accommodation (eastbound only),
Tel: 0800 555300.

Hartshead Moor
Junctions 25–26
Tel: 01274 876584
Welcome Break
Granary Restaurant. Fast food (Ltd). Shop.
Breakdown and repair facilities.
Accommodation (eastbound only), Tel:
01274 851706.

Ferrybridge
Off junction 33
Tel: 01977 672767
Granada
Restaurant. Fast food. Shop.
Accommodation, Tel: 0800 555300.

M27

Cadnam – Portsmouth

Rownhams
Junctions 3–4
Tel: 01703 734480
Roadchef.

Restaurant (westbound). Shop (Ltd). LPG
fuel available. Accommodation,
Tel: 01703 741144.

A1(M)

Scotch Corner – Tyneside

Durham
Off junction 61 (A177)
Tel: 0191 377 9222
Roadchef
Restaurant. Fast Food. Shop.
Accommodation, Tel: 0191 377 9222.

Washington
Junctions 64–65
Tel: 0191 410 3436
Granada
Restaurant. Shop. Accommodation
(southbound only), Tel: 0800 555300.

M40

London – Birmingham

Cherwell Valley
Junction 10
Tel: 01869 346060
Granada
Restaurant. Fast food – Burger King. Shop.
Breakdown and repair facilities.
Accommodation, Tel: 0800 555 300.

Warwick
Junctions 12–13
Tel: 01926 651681
Welcome Break
Restaurant. Shop. Accommodation,
Tel: 0800 850950.

M8

Edinburgh – Bishopton

Harthill
Junctions 4–5
Tel: 01501 751791
Roadchef
Restaurant. Shop.

M9

Edinburgh – Dunblane

Stirling
Off junction 9
Tel: 01786 813614
Granada
Restaurant. Shop. Accommodation,
Tel: 0800 555300.

M11

Birchanger Green
Off junction 8
Tel: 01279 653388
Welcome Break
Restuarant. Fast food. Granary.
Accomodation, Tel: 0800 850950.

M42

Bromsgrove – Measham

Tamworth
Off junction 10
Tel: 01827 260120
Granada
Restaurant. Shop. Accommodation,
Tel: 0800 555300.

M74

Glasgow – Abington
Bothwell
Junctions 4–5
Southbound only
Tel: 01698 854123
Roadchef
Restaurant. Fast food. Shop.

Hamilton
Junctions 5–6
Northbound only
Tel: 01698 282176
Roadchef
Restaurant. Shop. Accommodation,
Tel: 01698 891904.

Happendon
Between junctions 11–12
(B7078)
Tel: 01555 851880
Cairn Lodge
Restaurant. Fast food. Shop.
Accommodation.

Abington
Junction 13
Tel: 01864 502637
Welcome Break
Granary. Restaurant (Ltd). Shop.
Accommodation, Tel: 01864 502782.

A74 (M)

Abington – Gretna

Gretna
Junctions A75–B6357
Tel: 01461 337567
Welcome Break
Granary. Restaurant (Ltd). Shop.
Accommodation, Tel: 0800 850950.

Annandale Water
Junction 16 Johnstonebridge
Blue Boar
Restuarant. Accomodation, Tel: 0800 741174.

M90

Forth Bridge – Perth

Kinross
Off junction 6
Tel: 01577 863123
Granada
Restaurant. Fast food – Burger King. Shop.
Accommodation, Tel: 0800 555300.

28
Traffic Area
Offices

Traffic Area	*Counties Covered*
North Western Hillcrest House 386 Harehills Lane Leeds LS9 6NF Tel 0113 283 3533	the Metropolitan Counties of Greater Manchester and Merseyside, the Counties of Cheshire, Clwyd, Cumbria, Gwynedd and Lancashire and Derbyshire.
North Eastern Hillcrest House 386 Harehills Lane Leeds LS9 6NF Tel 0113 283 3533	the Metropolitan Counties of South Yorkshire, Tyne and Wear and West Yorkshire, the Counties of Cleveland, Durham, Humberside, Northumberland, North Yorkshire and Nottinghamshire.
West Midland Cumberland House 200 Broad Street Birmingham B15 1TD Tel 0121 608 1000	the Metropolitan County of West Midlands, the Counties of Hereford and Worcester, Shropshire,

Staffordshire and
Warwickshire.

Eastern
Terrington House
13–15 Hills Road
Cambridge
CB2 1NP
Tel 01223 358922

the Counties of
Bedfordshire,
Cambridgeshire,
Leicestershire,
Lincolnshire, Norfolk,
Northamptonshire and
Suffolk, the County of
Essex except the
Districts of Basildon,
Brentwood, Epping
Forest and Harlow and
the Borough of
Thurrock.

**South Eastern
and
Metropolitan**
Ivy House
3 Ivy Terrace
Eastbourne
BN21 4QT
Tel 01323 451400

the administrative area
of Greater London, the
Counties of East
Sussex, West Sussex
and the County of
Kent.

Scottish
J Floor
3 Lady Lawson
    Street
Edinburgh
EH3 9SE
Tel 0131 529 8500

Scotland

29
Police, Motoring and Weather Telephone Service Numbers

National And European Road Works And
  Weather Conditions
Police

NATIONAL AND EUROPEAN ROAD WORKS AND WEATHER CONDITIONS

Information on major road works and
weather conditions across the country is
available on the following telephone
numbers. This information has been supplied
by the Automobile Association.

National motorways 0336 401110
West Country 0336 401111
Wales 0336 401112
Midlands 0336 401113
East Anglia 0336 401114
North-west England 0336 401115
North-east England 0336 401116
Scotland 0336 401117
Northern Ireland 0336 401118
South-east England
    Central London (inside North/South
      Circulars) 0336 401122

Motorways/roads between the M4 and M1
0336 401123
Motorways/roads between M23 and M4
0336 401126
M25 London Orbital only 0336 401127
Continental Roadwatch 0336 401904
London Weather Centre (detailed enquiries)
0171 242 3663
Latest national forecast 0336 401130
European weather forecasts 0336 401105

POLICE

There are 12 police traffic area offices in the
country and in addition to being the issuing
authorities for LGV licences, they also keep
lists of police-approved recovery services in
their area and are often able to dispense
other useful information. Bear in mind that
the police can usually give reports on traffic
and weather conditions that are both more
up to the minute and less sensational than
those from almost any other source. It is
frequently well worth while phoning ahead
along your route to local police stations, to
enquire about road conditions. It is as well to
remember, however, that police stations are
often listed in the telephone directory under
the Police Force they belong to, rather than
under the country or town that they are in.

30
Breakdown and Recovery Services

The following list of companies operating recovery services has been adapted from material used in the *ABC Freight Guide*, published by Centaur Publications.

BEDFORDSHIRE

FLITWICK
Robson's Distribution
Station Approach, off Steppingley Road, Flitwick, Bedford MK45 1AL
Tel: 01525 718787 Fax: 01525 716177
Membership: FTA

BERKSHIRE

NEWBURY
Chartatruc Transport Ltd
Unit 11, Arnhem Road, Newbury, Berkshire RG14 5RU
Tel: 01635 30573 Fax: 01635 35403
Membership: BSI 5750, ISO 9000

CAMBRIDGESHIRE

STAPLEFORD
Welch's Transport Ltd
Granta Terrace, Stapleford, Cambridgeshire CB2 5DL
Tel: 01223 843011 Fax: 01223 843979
Membership: TA
Maximum lift: 40 tonnes

Maximum tow: 40 tonnes
Emergency tel no: 01223 892096

CHESHIRE

MIDDLEWICH
Bennion Transport & Commercial Vehicle Repairers
Brook Lane Industrial Estate, Middlewich, Cheshire CW10 OJG
Tel: 01606 835811 Fax: 01606 836228
Additional Services: Local breakdown fitter attends in van

DERBYSHIRE
Sam Longson Ltd
Town End Garage, Chapel-en-le-Frith, High Peak, Derbyshire SK23 0PF
Tel: 01298 812301 Fax: 01298 815013 Telex: 669077
Membership: FTA, IRTE, CIT
Maximum lift: 20 tons
Maximum tow: 40 tons
Additional services: Roadside repairs, full engineering services on site. Hazchem trained staff/specialist equipment, ERF dealer, spares available

CLWYD

WREXHAM
Clwyd Transport Services Ltd
21 Grosvenor Road, Wrexham
LL11 1BT
Tel: 01978 290 203 Fax: 01978 290 301
Emergency tel. no: 01836 573 891

CORNWALL

ST AUSTELL
ECC International Ltd
John Keay House, St Austell, Cornwall
PL25 4DJ
Tel: 01726 824444 Fax: 01726 623019

Membership: FTA, TA
Maximum lift: Any
Maximum tow: Any
Emergency tel no: 01726 858810 (24 hours)

N J Grose Ltd
Bucklers Lane, Holmbush, St. Austell,
Cornwall PL25 3JN
Tel: 01726 74551 Fax: 01726 76224
Additional services: Pre-MOT and service
maintenance contracts

DERBYSHIRE

MATLOCK
Matlock Transport Company Ltd
Cotehillock Garage, Northwood Lane,
Darley Dale, Matlock, Derbyshire DE4 2HQ
Tel: 01629 733357 Fax: 01629 732176
Maximum lift: 38 tonnes
Maximum tow: 38 tonnes
Additional services: RHA Recovery
Membership

Biagellia Transport Ltd
187 The Hill, Cromford, Nr Matlock,
Derbyshire DE4 3QS
Tel: 01629 825000 Fax: 01629 825662
Maximum lift: 10 tonnes
Maximum tow: 30 tonnes
Additional services: Contract maintenance,
vehicle painting, equipment fitted

DEVON

EXETER
Maddern Transport Limited
Kestrel Way, Sowton Industrial Estate,
Exeter, Devon EX2 7JQ
Tel: 01392 250591
Fax: 01392 420087
Membership: Institute of Transport
Administration

AVRO membership: Yes
Maximum lift: 20 tons
Maximum tow: 38 tons
Additional services: Replacement vehicles
Emergency tel no: 01392 53152 or
01392 79540 or 01392 431741

CO. DURHAM

BISHOP AUCKLAND
Davidsons of Coundon
West House, Coundon, Bishop Auckland,
Co. Durham DL14 8PS
Tel: 01388 602688 Fax: 01388 602688
Membership: TA
Additional services: MOT repairs

DARLINGTON
T F Corner Ltd
Arlaw Banks, Winston, Darlington,
Co. Durham DL2 3PX
Tel: 01833 627252 Fax: 01833 627532
AVRO membership: Yes

George Allinson Transport Ltd
Faverdale North, Faverdale Industrial
Estate, Darlington, Co. Durham DL3 OPH
Tel: 01325 461 241 Fax: 01325 464 651
Maximum lift: 10 tonnes, heavy list
Maximum tow: 20 tonnes
Additional services: Up to 30 tonnes in depot

O. Stiller (Transport) Ltd
Boeing Way, Preston Farm Industrial Estate,
Stockton on Tees, Cleveland, TS18 3TE
Tel: 01642 607777/01642 607711
Membership: FTA
Maximum lift: Trailer
Emergency tel no: 01642 607777

L Swindells & Sons
The Goods Depot, Piercebridge, Darlington,
Co. Durham DL2 3TS

Tel: 01325 374506/374282 Fax: 01325 374665
Maximum lift: 18 tonnes
Maximum tow: 20–22 tonnes
Emergency tel no: 01325 286669

NEWTON AYCLIFFE
Fracks Transport Ltd
Groat Drive, Aycliffe Industrial Estate,
Newton Aycliffe, Co. Durham DL5 6HY
Tel: 01325 300440 Fax: 01325 317017
Maximum lift: 8 tons
Maximum tow: 38 tons

ESSEX

CHELMSFORD
Heavyhaul (Chelmsford) Ltd
Landview House, Cooksmill Green, Ongar
Road West, Chelmsford, Essex CM1 3SR
Tel: 01245 248577 Fax: 01245 248213
AVRO membership: Yes
Maximum lift: 12 tonnes
Maximum tow: 150 tonnes
Additional services: Winch trucks,
HIAB trucks, cranes
Emergency tel no: 01245 248705

COLCHESTER
N C Cammack and Son Ltd
Tyburn Hill, Colchester Road, White Colne,
Colchester, Essex CO6 2PS
Tel: 01787 222001 Fax: 01787 223230
AVRO membership: Yes
Maximum lift: 10 tonnes
Maximum tow: 60 tonnes
Additional services: Commercial vehicle
repairs
Emergency tel no: 01787 223121

GIDEA PARK
Trouble Free Truckin'
117A Belgrave Avenue, Gidea Park, Essex
RM2 6PS

Maximum lift: 50 tonnes
Maximum tow: Any

SOUTH GLAMORGAN

CARDIFF
D A Dyer & Son Ltd
Unit 6, Clydesmuir Industrial Estate,
Tremorfa, Cardiff, South Glamorgan
CF2 2QS
Tel: 01222 496 502 Fax: 01222 499 262
AVRO membership: Yes
Maximum lift: 40 tonnes
Maximum tow: 75 tonnes
Additional services: Airbags
Emergency tel no: 01222 496 502

GLOUCESTERSHIRE

GLOUCESTER
Joseph Rice Transport
26 Hempsted Lane, Gloucester, Glos
GL2 5JF
Tel: 01452 527473 Fax: 01452 300456
Membership: FTA, IRTE
Maximum tow: 24 tonnes
Emergency tel no: 01452 522563

GRAMPIAN

ABERDEEN
Sandy Bruce Trucking
Blackdog Industrial Centre, Murcar,
Aberdeen, Grampian AB28 8BT
Tel: 01224 824936 Fax: 01224 820945

GWENT

NEWPORT
Coryton Motor Co Ltd
29 Malpas Road, Newport, Gwent
NP9 6WA
Tel: 01633 822456
Membership: AVRO

HAMPSHIRE

BASINGSTOKE
Wheatsheaf Garage
North Waltham, Basingstoke, Hants
RG25 2BB
Tel: 01256 397254 Fax: 01256 397119
AVRO Membership: Yes
Maximum lift: 38 tonnes
Maximum tow: 38 tonnes
Additional services: Air cushions
Emergency tel no: 01256 397254

PORTSMOUTH
Hendy Ford Cosham Ltd
Southampton Road, Cosham, Portsmouth,
Hampshire PO6 4RW
Tel: 01705 322900 Fax: 01705 321725
Membership: FTA
Maximum lift: 40 tonnes
Maximum tow: 40 tonnes
Additional services: Roadside recovery,
mobile workshop, vehicle washing, MoT,
vehicle sales, tachograph testing station, pre-
ministry testing, light commercial tyres
Emergency tel no: 01705 370944

SOUTHAMPTON
Meachers Transport
Unit 19, Mauretania Road, Nursling
Industrial Estate, Southampton, Hampshire
SO16 0YS
Tel: 01703 739999 Fax: 01703 730019
Maximum lift: 38 tonnes
Maximum tow: 38 tonnes
Additional services: Repairs
Emergency tel no: 01703 741010

HIGHLAND

INVERNESS
Highlands & Islands Enterprise (Government Agency)

Bridge House, 20 Bridge Street, Inverness,
Highland IV1 1QR
Tel: 01463 234171
Membership: FTA
Maximum lift: Lorry weight
Maximum tow: Empty low
Additional services: As required

WICKS
Alexander Building, Scrabster Business Park,
Scrabster KW14 7UG
Tel: 01955 602381
Membership: FTA
AVRO membership: Yes

HUMBERSIDE

GRIMSBY
Exxtor Group Limited
P.O. Box 40, Manby Road, Immingham,
North East Lincolnshire DN40 3EG
Tel:01469 571711 Fax: 01469 573504
Membership: BS5750 Certification

Humberside Transport
Estate Road 2, South Humberside Industrial
Estate, North East Lincolnshire
DN31 2TE
Tel: 01472 346805 Fax: 01472 242190
Telex: 52409
AVRO membership: Yes
Maximum lift: 8 tonnes
Maximum tow: 3 tonnes
Additional services: 24 hour local cover
Emergency tel no: 01472 252457

HULL
I J Blakey Haulage Co Ltd
Fleet House, Woodhouse Street, Hedon
Road, Hull, North Humberside HU9 1RJ
Tel: 01482 324347 Fax: 01482 216489
Maximum lift: 12 tons
Maximum tow: 60 tons

Additional services: Comprehensive garage
facilities
Emergency tel no: 01482 503436

David C Tully
Unit B, Shirethorn Business Park, Thorn
Firm, Cleveland Street, Hull, East Yorkshire
HU8 7BA
Tel: 01482 585751
Additional services: Roadside assistance
Emergency tel no: 01482 23490

KENT

MAIDSTONE
Lenham Storage Co Ltd
Ham Lane Lenham, Maidstone, Kent
ME17 2LH
Tel: 01622 858 441 Fax: 01622 850 469 Telex:
96225
Membership: FTA
Maximum lift: 38 tonnes

ROCHESTER
Squires & Knight
82 King Street, Rochester, Kent ME1 1YD
Tel: 01634 841651/213 Fax: 01634 829494

WEST MALLING
D H Yates & Sons (Transport) Ltd
Brickfields Depot, London Road, West
Malling, Kent ME19 5AG
Tel: 01622 673980
Membership: AVRO
Maximum lift: 3 ton
Maximum tow: 20 ton
Additional services: Repair, truck rental
Emergency tel no: 01732 848516

LANCASHIRE

ARNSIDE
G A Pearson General Haulage

Trafalgar Garage, Ashley Road, Carnforth,
Lancashire
LA5 OH3
Tel: 01524 761481 Fax: 01524 761481
Additional services: Light and heavy
commercial vehicles

PRESTON
J M Sanderson International Horse Transport
Dunroamin Farm, Hollins Lane, Forton,
Preston, Lancashire PR3 OAA
Tel: 01524 791251

LEICESTERSHIRE

LEICESTER
A M Widdowson & Son Ltd
The Mill Lane Industrial Estate, Kirby
Road, Glenfield, Leicester, Leicestershire
LE3 8DX
Tel: 0116 2312620 Fax: 0116 2313303
AVRO membership: Yes
Maximum lift: 38 tonnes
Maximum tow: 150 tonnes
Additional services: Full maintenance and
repair service
Emergency tel no: 0116 2877345

LUBENHAM
Badger Bros
109 Main Street, Lubenham, Leicestershire
LE16 9TG
Tel: 01858 466984 Fax: 01858 433735
AVRO membership: Yes
Maximum lift: 30 tonnes
Maximum tow: 100 tonnes
Additional services: Roadside assistance,
winching on/off road, Hazchem trained staff,
low loader workshop facilities
Emergency tel no: 01858 466984

LINCOLNSHIRE

CAISTOR
H C Wright & Sons
The Mill, Whitegate Hill, Caistor,
Lincolnshire LN7 6SW
Tel: 01472 851461 Fax: 01472 851952
Emergency tel no: 01472 852469

SPALDING
Grantham's Recovery Ltd
6 Main Road, Quadring, Spalding,
Lincolnshire PE11 4PS
Tel: 01775 840452 Fax: 01775 840506
Maximum lift: 50 tonnes
Maximum tow: 100 tonnes
Additional services: Crane hire, air cushions,
workshop facilities
Emergency tel no: 01775 760446 and
01205 85479

LONDON

D J C Distribution Ltd
61–81 Eastmoor Street, Charlton, London
SE7 8LX
Tel: 0181 858 8601 Fax: 0181 858 1258
Membership: NAWK
Maximum tow: 10 tonnes
Additional services: All types of repairs

Otway & Golder Ltd
40a Jasmine Grove, Penge, London
SE20 8JP
Tel: 0181 778 7047 Fax: 0181 676 8034
Membership: FTA
Maximum lift: Unlimited
Maximum tow: Unlimited
Emergency tel no: 0181 656 5526

Twyford Commercials Ltd
1 Chambers Street, London SE16 4XQ
Tel: 0171 237 1566 Fax: 0171 237 4391

AVRO membership: Yes
Maximum lift: 10 tonnes
Maximum tow: 60 tonnes
Additional services: Crane hire, mobile welding
Emergency tel no: 0171 237 1566

MERSEYSIDE

LIVERPOOL
D T & P Chadwick Ltd
North End Garage, Gerrards Lane,
Halewood, Liverpool, Merseyside L26 5QA
Tel: 0151 488 6888 Fax: 0151 487 9489

Gobowen Transport Ltd
E W M House, 107 Rimrose Road,
Liverpool, Merseyside L20 4HN
Tel: 0151 933 9900 Fax: 0151 933 7623

NORTHAMPTONSHIRE .

KETTERING
Pink & Jones Ltd
Riley Road, Telford Way Industrial Estate,
Kettering, Northamptonshire
NN6 8UW
Tel: 01536 512019 Fax: 01536 410584
Maximum tow: DK

NORTHAMPTON
Airflow Streamlines Ltd
Hopping Hill, New Duston, Northampton,
Northamptonshire NN5 6PD
Tel: 01604 581121 Fax: 01604 595885
AVRO membership: Yes
Maximum lift: 8 tons
Maximum tow: 38 tons
Additional services: Iveco/Ford agent, Lucas
Kienze tacho and speed limiters
Emergency tel no: 01604 581121

NOTTINGHAMSHIRE

LENTON
J Stirland Ltd
Willow Road, Lenton, Nottinghamshire
NG7 2SN
Tel: 0115 9781365 Fax: 0115 9423127
Additional services: Minor repairs

NEWARK
Griffins of Newark Limited
Main Street, Normanton-on-Trent, Newark,
Nottinghamshire NG23 4RQ
Tel: 01636 821600 Fax: 01636 22092
Additional services: Roadside as required

OXFORDSHIRE

THAME
Leyland Daf
Eastern Bypass, Thame, Oxfordshire
OX9 3FB
Tel: 01844 261 111 Fax: 01844 217 234
Additional services: Leyland DAF AID
Emergency tel no: 0800 919 395

SHROPSHIRE

OSWESTRY
G & R Cadwallader Ltd
Maesbury Road, Oswestry, Shropshire
SY10 8NJ
Tel: 01691 655721 (Continental),
01691 653156 (UK) Fax: 01691 658085
Telex: 35306
Membership: RHA International Group

SOMERSET

TAUNTON
South West Motors Ltd
Cornishway North, Galmington Trading
Estate, Taunton, Somerset TA1 5LY

Tel: 01823 327805 Fax: 01823 321791
AVRO membership: Yes
Maximum lift: 10 tons
Maximum tow: 77 tons
Additional services: 24 hour car and van
recovery
Emergency tel no: 01823 327805 (24 hour)

STAFFORDSHIRE

STAFFORD
Pasturefields Recovery Services
Drummond Road, Astonfields Industrial
Estate, Stafford, Staffordshire ST16 3HJ
Tel: 01785 254495 Fax: 01785 226413
Membership: AVRO, RMIF
AVRO membership: Yes
Maximum lift: 8 tonnes
Maximum tow: 100 tonnes
Emergency tel no: 01785 54495

STRATHCLYDE

GLASGOW
John Smith Transport Contractor Ltd
85 Clydeholm Road, Clydeside Industrial
Estate, Whiteinch, Glasgow G14 0SE
Tel: 0141 954 8071 Fax: 0141 954 1877
Membership: FTA, NAWK
Maximum tow: Unladen
Emergency tel no: 0141 954 8071

TAYSIDE

PERTH
McLaughlan Transport
Lower Harbour, Friarton Road, Perth,
Tayside PH2 8BB
Tel: 01738 634321 Fax: 01738 30060
Membership: TFR

WEST MIDLANDS

WALSALL
S Jones Transport Service Group
Anglian Road, Aldridge, Walsall, West
Midlands WS9 8ET
Tel: 01922 54411 Fax: 01922 52709 Telex:
338559
Maximum lift: 10 tons
Maximum tow: 38 tons
Additional services: Recovery
Emergency tel no: 01922 743783

WILLENHALL
Joseph Foulkes (Wednesfield) Ltd
Meachells Lane, Willenhall, West Midlands
WV13 3RP
Tel: 01902 368521 Fax: 01902 635356 Telex:
336014
Membership: FTA

YORKSHIRE

BRIDLINGTON
Kidds Services
1 & 2 Bessingby Way, Bessingby Industrial
Estate, Bridlington, East Yorkshire YO16
4SJ
Tel: 01262 672842/679589/676618 Fax:
01262 603312
Membership: British Association of
Removers, National Association of
Warehousekeepers
Additional services: HGV garage
Emergency tel no: 01262 672842

WEST YORKSHIRE

LEEDS
**National Breakdown Commercial Recovery
Ltd**
P.O. Box 300, Cote Lane, Leeds, West
Yorkshire LS99 2LZ
Tel: 0800 400600 Fax: 0113 273111

Membership: FTA
AVRO membership: Yes
Maximum lift: 44 tonnes
Maximum tow: 44 tonnes
Additional services: European service,
trailers, refrigeration, tail lifts
Emergency tel no: On application

31
Return Loads
Section

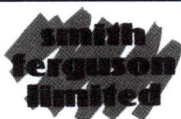

32
Useful Conversions

LENGTH

1 millimetre	– 1000 micrometres	– 0.0394 inch
1 centimetre	– 10 millimetres	– 0.3937 inch
1 metre	– 100 centimetres	– 1.0936 yards
1 kilometre	– 1000 metres	– 0.6214 mile
1 inch		– 2.54 centimetres
1 foot	– 12 inches	– 30.48 centimetres
1 yard	– 36 inches	– 0.9144 meter
1 mile	– 1760 yards	– 1.6093 kilometres

AREA

1 sq metre	– 10,000 sq centimetres	– 1.196 sq yards
1 hectare	– 10,000 sq metres	– 2.4711 acres
1 sq kilometre	– 100 hectares	– 0.3861 sq mile
1 sq foot	– 144 sq inches	– 0.0929 sq metre
1 sq yard	– 9 sq feet	– 0.8361 sq metre
1 acre	– 4840 sq yards	– 4046.9 sq metres

CAPACITY

1 cu decimetre	– 1000 cu centimetres	– 0.0353 cu foot
1 cu metre	– 1000 cu decimetres	– 1.3080 cu yards
1 litre	– 1 cu decimetre	– 0.22 gallon
1 cu yard	– 27 cu feet	– 0.7646 cu metre
1 pint	– 4 gills	– 0.5683 litre
1 gallon	– 8 pints	– 4.5461 litres

WEIGHT

1 gram	– 1000 milligrams	– 0.0353 ounce
1 kilogram	– 1000 grams	– 2.2046 pounds
1 tonne	– 1000 kilograms	– 0.9842 ton
1 ounce	– 437.5 grains	– 28.35 grams
1 pound	– 16 ounces	– 0.4536 kilogram
1 stone	– 14 pounds	– 6.35 kilograms

Appendix
Dangerous Goods
Driver Training
Centres

ENGLAND

AVON

Nailsea

National ADR Training Consultancy
6 Goss Barton
Nailsea, nr Bristol BS19 2XD
Tel: 01275 857920

Portbury

Training Force
Regional Freight Centre
Portbury
Bristol BS20 9XX
Tel: 01275 374401 *(Mobile courses available)*

BERKSHIRE

Slough

Fullers Transport Training Services
724 Dundee Road
Slough, SL1 4JQ
Tel: 01753 530829

Sunninghill

Nationwide Transport Training
Crossman House

9 High Street
Sunninghill SL5 9NQ
Tel: 01344 875481

CAMBRIDGESHIRE

Cambridge

East Anglian Driver Training
Granta Terrace
Stapleford, Cambridge CB2 5DJ
Tel: 01223 845342

Peterborough

Mid-Anglia HGV Training Ltd
138 Peterborough Road
Whittlesey, Peterborough
Cambs PE7 1PD
Tel: 01733 202156

CHESHIRE

Ellesmere Port

Calor Gas Ltd
Dockyard Road
Ellesmere Port
South Wirral L65 4EG
Tel: (0151) 355 3700

Runcorn

Chemfreight Training Ltd
Cormorant Drive
Runcorn, Cheshire WA7 4UD
Tel: 01928 580505

Warrington

North Cheshire Training Association Ltd
Unit 11/4, Pallatine Industrial Estate
Causeway Avenue
Warrington WA4 6QQ
Tel: 01925 658342

CLEVELAND

Middlesbrough

Centre for Industrial and Commercial
Training
Hamilton House, Cargo Fleet
Middlesbrough, Cleveland TS3 8DJ
Tel: 01642 222821

Teesside Training Enterprise – Management
and Technical Training
PO Box 36
Wilton Training Centre
Wilton Site, Middlesbrough, TS6 8YX
Tel: 01642 433295

COUNTY DURHAM

Langley Moor

Firlands Training Ltd
Little Burn Industrial Estate
Langley Moor DH7 8HG
Tel: 0191 378 0601

CUMBRIA

West Cumbria Training Ltd
23B Solway Estate
Maryport CA15 8NF
Tel: 01900 814560

Penrith

Cumbria Fire Service Training Centre
Bridge Lane
Penrith CA11 8HY
Tel: 01900 822503

DEVON

Exeter

Trans-Plant/Mastertrain

Exeter Livestock Centre
Matford Park Road
Exeter EX2 8FD
Tel: 01392 426242

DORSET

Bournemouth

GB Express Ltd
Elliot Road
Bournemouth BH11 8JR
Tel: 01202 580707

Poole

Wessex Transport Training Ltd
59 Old Wareham Road
Parkestone, Poole BH12 4QN
Tel: 01202 717881

EAST SUSSEX

Hastings

Driver Training Centre
2 Alexandra Parade, Park Ave
Hastings TN34 2PQ
Tel: 01424 432200

ESSEX

Colchester

TDT Management Services Ltd
London Road
Kelvedon, Colchester
Essex CO5 9AU
Tel: 01376 570863

VAS (Vocational and Safety) Training Ltd
4 Churchwell Avenue
CO5 9HN
Tel: 01206 213443

Grays

ADR Modular Training
Lyndale Court
London Road
West Thurrock
Grays, Essex RM20 3BJ
Tel: 01708 861339

Harlow

P & O Distribution Ltd
River Way
Harlow, Essex CM20 2HB
Tel: 01279 418341

West Thurrock

Wright Training Services Ltd
Truckworld
Oliver Road
West Thurrock
Essex RM20 3ED
Tel: 01708 867564

GREATER MANCHESTER

Bolton

Red Rose Training
Europa Way
Stoneclough, Radcliffe M26 9GG
Tel: 01204 862999

Manchester

Air Products (UK) Ltd
Manchester Road, Carrington
Manchester M31 4GT
Tel: (0161) 775 4381

Hargreaves Training Services Ltd
Block A, The Court
Kestrel Road
Trafford Park, Manchester M17 1SF
Tel: (0161) 872 7916

Manchester Training
Greengate
Middleton
M24 1RU
Tel: (0161) 653 5767

P&O Roadways Ltd
Hibernia Way
Barton Dock Road
Manchester M32 OYJ
Tel: (0161) 911 5300

HEREFORD

Hereford

Three Counties Training Services Ltd
12A The Cattle Market
Hereford HR4 9HX
Tel: 01432 269126

HERTFORDSHIRE

St Albans

West Herts Training Ltd
VER House
Park Industrial Estate
Frogmore, St Albans AL2 2WH
Tel: 01727 873878

KENT

Gillingham

Carlton Training Services
Courteney Road, Gillingham ME8 0RT
Tel: 01634 372380

Tunbridge Wells

Freight Transport Association
Hermes House, St John's Road
Tunbridge Wells TN4 9UZ
Tel: 01892 526171

LANCASHIRE

Bury

Linkman Tankers
Moss Hall Road
Lancashire BL9 8RW
Tel: (0161) 764 6464

LEICESTERSHIRE

Hinckley

Garage and Transport Training Ltd
Jacknell Road
Dodwells Bridge Industrial Estate
Hinckley LE10 3BS
Tel: 01455 251516

Leicester

J Coates (HGV Services) Ltd
46/50 Great Central Street
Leicester LE1 4NF
Tel: 0116 262 6037

Stoney Stanton

Calor Gas Ltd
Occupation Road
Stoney Stanton LE9 6JJ
Tel: 01455 272273

LONDON

Kamillionex Communications Ltd
K Training
Empire House
Empire Way
Wembley
London HA9 0EW
Tel: 0181 970 2132

MERSEYSIDE

Liverpool

ACL HGV Training Ltd
Goras Road
Kirby Industrial Estate North,
Kirby, Liverpool L33 7XS
Tel: 0151 549 2543

Trafalgar Training Centre
Bridge House
South Trafalgar Dock
Liverpool L3 0AG
Tel: 0151 236 4592

St Helens

Sutton & Sons (St. Helens) Ltd
Sutton Heath
St Helens, Merseyside WA9 5BW
Tel: 01744 811611

NORFOLK

Norwich

Norfolk Training Services
Harford Centre, Hall Road
Norwich NR4 6DG
Tel: 01603 259900

NORTHANTS

Northampton

TNT Express Worldwide UK Ltd
Training Dept
South Portway Close
Round Spinney Ind. Estate
Northampton NN3 8RB
Tel: 01604 671234

NORTH HUMBERSIDE

Hull

Training for Industry (Humberside) Ltd
Willow House
Clay Street, Hull HU8 8HA
Tel: 01482 223520

NOTTINGHAMSHIRE

Beeston

Firlands Training Ltd
GPT Business Park
Unit C2
Technology Drive
Beeston NG9 2ND
Tel: 0115 922 1225

Kimberley

Trent Transport Training
The Watson Centre
Artic Way, Kimberley
Nottingham NG16 2HS
Tel: 0115 938 4982

Kirkby-in-Ashfield

B. Taylor & Sons Ltd
Byron Avenue
Low Moor Ind. Estate
Kirkby-in-Ashfield NG17 7LA
Tel: 01623 759595

Retford

Haygarth Training Centre
Haygarth House
Babworth, Retford DN22 8ES
Tel: 01777 708681

Worksop

North Notts Training Group
Claylands Avenue
Dukeries Industrial Estate
Worksop S8I 7DJ
Tel: 01909 475745

OXFORDSHIRE

Abingdon

Thames Valley Training
Blakes Oak Farm
Lodge Hill
Abingdon OX14 2JD
Tel: 01235 530085

Didcot

Air Products (UK) Ltd
Harrier Road
Hawkesworth
Didcot OX11 7PL
Tel: 01235 510771

Enstone

Rollright School of Transport
Enstone Airfield Complex
Enstone OX7 4NP
Tel: 01608 677618

Wallingford

Ridgeway International Ltd
69 High Street
Wallingford OX10 0BX
Tel: 01491 839780

SOMERSET

Bridgewater

Wessex Vehicle Training
Suprema Estate
Edington, Bridgewater TA7 9LF
Tel: 01278 723113

Shepton Mallet

Skilltrain Associates
17 Town Street
Shepton Mallet BA4 5BE
Tel: 01749 345681

Taunton

Friendberry Ltd
Stogumber, Taunton TA4 3TP
Tel: 01984 656310

SOUTH HUMBERSIDE

Grimsby

Management and Industrial Training Ltd
260 Macaullay Street
Grimsby, South Humberside DN31 2EY
Tel: 01472 329329

SOUTH YORKSHIRE

Doncaster

Doncaster Rotherham and District
   Motor Trades Group Training Assn
Rands Lane Industrial Estate
Rands Lane, Armthorpe
Doncaster DN3 3DY
Tel: 01302 832831

STAFFORDSHIRE

Leek

Transed Associates
Springbuck House
Leekbrook Ind. Estate
Cheadle Road
Leek ST13 7AP
Tel: 01538 381313
(Contact Mrs M Wood)

Stafford

Leigh Environmental Ltd
Dunstan Hall
Dunstan, Stafford ST18 9AB
Tel: 01785 712666

SUFFOLK

Felixstowe

P&O Transport Training
Sub Station Road
Felixstowe, Suffolk IP11 8SE
Tel: 01394 674139

Securicor Omega Container Logistics
2 Nicholas Road
Felixstowe IP11 8XH
Tel: 01394 673443

Ipswich

RTT Training Services Ltd
Gyppeswyk Hall
Gyppeswyk Avenue, Ipswich
Suffolk IP2 9AF
Tel: 01473 602424

SURREY

Bookham

EP Training Services Ltd
The Old Library
Lower Shott, Great Bookham
Surrey KT23 4LR
Tel: 01372 450800

Guildford

The British Oxygen Co (BOC) Ltd
The Priestly Centre
10 Priestly Road
The Surrey Research Park
Guildford GU2 5XY
Tel: 01483 579857

TYNE AND WEAR

Gateshead

Colin Hall Assessment and Training
  Services
26 Mullen Gardens
Wallsend, NE28 9E2
Tel: (0191) 263 7507

Van Hee Transport Ltd
Training Dept, William Street
Felling, Gateshead
Tyne and Wear NE10 0JP
Tel: (0191) 438 2512

Hebburn

Tyne and Wear ADR Training
Station Road
Hebburn NE31 1NY
Tel: (0191) 430 0505

Newcastle upon Tyne

Tyneside Training Services Ltd
Airport Industrial Estate
Kingston Park, Kenton
Newcastle upon Tyne NE3 2EF
Tel: (0191) 286 2919

WARWICKSHIRE

Warwick

CONOCO Ltd
CONOCO Centre
Warwick Technology Park
Gallows Hill, Warwick CV34 6DA
Tel: 01926 404000

WEST MIDLANDS

Aldridge

Westgate Training

Gainsborough House
Brickyard Road
Aldridge
Walsall WS9 8SR
Tel: 01922 55514

Kingswinford

West Midlands Training Group Ltd
Dudley Road, Kingswinford
West Midlands DY6 8BS
Tel: 01384 401199

Wolverhampton

BOC Ltd, *(Training Centre)*
Lower Walsall Street
Wolverhampton
West Midlands WV1 2EP
Tel: 01902 870999

WEST SUSSEX

Crawley

Shell UK Downstream (SUKD)
Colas Ltd
Rowfant, Crawley RH10 4NF
Tel: 01342 711087

WEST YORKSHIRE

Batley

Haz Training Services (contact Mrs. C A P
Walker)
10 Wren Hill
Woodlands Road, Batley WF17 0QL
Tel: 01924 440935

Bradford

Ellis & Everard (UK) Ltd
46 Peckover Street
Bradford BD1 5BD
Tel: 01274 377000

West Yorkshire Training Group
420 Tong Street,
Bradford, West Yorkshire BD4 6LP
Tel: 01274 689814

Leeds

ASC Training Services
ASC House, 54 North Street
Leeds LS2 7PN
Tel: 0113 245 3480

Hargreaves Training Services Ltd
Unit 3, Glover Way
Parkside Industrial Estate
Leeds, West Yorkshire LS11 5JP
Tel: 0113 270 1188

Wakefield

Sandal Business Services
80 Limepit Lane
Stanley, Wakefield
West Yorkshire WF3 4DF
Tel: 01924 820078

WILTSHIRE

Devizes

Wiltshire Transport Training Ltd
Hopton Industrial Estate
London Road, Devizes SN10 2EX
Tel: 01380 723712

WALES

CLWYD

Sandycroft

Cameon Ltd
Allan Morris Ltd
Factory Road, Sandycroft
Deeside CH5 2QJ
Tel: 01244 533320

Wrexham

Gatewen Training
Gatewen, New Broughton
Wrexham, Clwyd LL11 6YA
Tel: 01978 720907

DYFED

Pembroke Dock

Mainport Training
Pembrokeshire Enterprise Centre
Kingswood
Pembroke Dock, Pembrokeshire SA72 4RS
Tel: 01646 684315

SOUTH GLAMORGAN

Barry

Barry Training Services
Holt Buildings
Powell Duffryn Way
No. 1 Dock, Barry CF62 5QS
Tel: 01446 739457

SCOTLAND

Aberdeen

EJS Transport Training
Unit One
Deemouth Centre
South Esplanade East
Torry
Aberdeen AB11 9PB
Tel: 01224 898946

J. Gilbert
Broadfield Road
Bridge of Don
Aberdeen AB23 8EE
Tel: 01224 825644

Bon-Accord Training Services
The Training Centre
Hill of Menie
Balmedie, Aberdeen AB23 8YD
Tel: 01358 742121

Scottish Offshore Training Association
Blackness Avenue
Altens, Aberdeen AB12 3PG
Tel: 01224 899707

Bonnybridge

LAGTA Central
Seabegs Road, Bonnybridge FK4 2AQ
Tel: 01324 812035

Coatbridge

LAGTA Ltd
7 Palcecraig Street
Shawhead, Coatbridge ML5 4RY
Tel: 01236 426171

Dundee

Tayside Training Company Ltd
Smeaton Road, Wester Gowdie
Dundee DD2 4UT
Tel: 01382 623261

Dundonald

Securicor Omega International
Olympic Complex
Drybridge Road
Dundonald, Ayrshire KA2 9BE
Tel: 01563 851212

Glasgow

Glasgow Training Group
120 Crowhill Road
Bishopsbriggs, Glasgow G64 IRP
Tel: (0141) 762 1461

Ritchies Training Centre Ltd
Hobden Street, Glasgow G21 4AQ
Tel: (0141) 557 2212

Stirling

Freight Transport Association
Hermes House, Melville Terrace
Stirling FK8 2ND
Tel: 01786 457500

Motherwell

Motherwell College
Dalzell Drive
Motherwell ML1 2DD
Tel: 01698 232323

NORTHERN IRELAND

Crumlin

Transport Training Services Ltd
15 Dundrod Road
Nutts Corner, Crumlin
Co Antrim BT29 4SS
Tel: 01232 825653

Index of Advertisers